LOCATING DIALECT IN DISCOURSE

LOCATING DIALECT IN DISCOURSE

The Language of Honest Men and Bonnie Lasses in Ayr

RONALD K. S. MACAULAY

New York   Oxford
OXFORD UNIVERSITY PRESS
1991

Oxford University Press

Oxford   New York   Toronto
Delhi   Bombay   Calcutta   Madras   Karachi
Petaling Jaya   Singapore   Hong Kong   Tokyo
Nairobi   Dar es Salaam   Cape Town
Melbourne   Auckland

and associated companies in
Berlin   Ibadan

Library of Congress Cataloging-in-Publication Data
Macaulay, Ronald K. S.
Locating dialect in discourse : the language of honest men
and bonnie lasses in Ayr / by Ronald K. S. Macaulay.
p. cm. — (Oxford studies in sociolinguistics)
Includes bibliographical references and index.
ISBN 0-19-506559-X
1. Scots language—Dialects—Scotland—Ayr.
2. Scots language—Social aspects—Scotland—Ayr.
3. Scots language—Discourse analysis.
4. Urban dialects—Scotland—Ayr.
5. Ayr (Scotland)—Social conditions.
I. Title.   II. Series.
PE2169.A97M33   1991
306.4'4'09411—dc20
90-45065

1 3 5 7 9 8 6 4 2

Printed in the United States of America
on acid-free paper

For Janet
my bonniest lass

Series Foreword

Sociolinguistics is the study of language in use. Its special focus, to be sure, is on the relationships between language and society, and its principal concerns address linguistic variation across social groups and across the range of communicative situations in which men and women deploy their verbal repertoires. In short, sociolinguists examine language as it is constructed and co-constructed, shaped and reshaped, in discourse.

While some linguists analyze the structure of sentences independent of their context (independent of who is speaking or writing and to whom, of what has preceded and is to follow, and of the setting and purposes of the discourse), sociolinguists investigate language as it is embedded in its social and situational contexts. It is not surprising that virtually all interest in language among people who are linguists also focuses on language in context, for the patterns of language in use reflect the intricacies of social structure and mirror the salient situational and strategic factors shaping discourse.

In offering a platform for studies of language use in communities around the globe, Oxford Studies in Sociolinguistics invites significant treatments of discourse and of social dialects and registers, whether oral, written, or signed, whether synchronic or diachronic. The series will host studies that are descriptive and theoretical, interpretive and analytical. While most volumes will report original research, a few will synthesize existing knowledge. All volumes will aim for a style accessible not only to linguists but to other social scientists and humanists interested in language use. Occasionally, if we are fortunate, a volume will appeal beyond scholars and students to those educated readers keenly interested in the everyday discourse of human affairs: for example, the discourse of doctors or lawyers trying to engage clients with their specialist tongues, or of women and men striving to grasp the sometimes baffling character of their shared interactions.

With *Locating Dialect in Discourse,* Oxford Studies in Sociolinguistics presents its first volume. Ronald Macaulay's impressive study provides a wealth of information about the speech patterns of twelve "honest men" and "bonnie lasses" of Ayr, Scotland. Broader in scope than most previous sociolinguistic studies, it offers detailed analysis of the phonological, syntactic, lexical, and discourse features used by two social groups in Ayr. Professor Macaulay's coverage of an unusually wide range of topics combines with his lively and engaging style to give this book a special appeal.

Because of great variation in the kinds of data gathered by sociolinguists in the past, few comparisons across communities can be made, as Professor Macaulay notes. By contrast, his study offers researchers a methodology that can be widely adopted, for such conversations as he recorded between "intimate strangers" in Ayr are available to skillful interviewers almost everywhere. Indeed, Professor Macaulay's findings offer eloquent testimony to the richness of his methodology, and *Locating Dialect in Discourse* offers other researchers an invitation—a challenge, even—to extend the analysis of sociolinguistic variation to conversation in communities far and wide. It is a worthy challenge, for only with a wider data base than now exists—and an abundance of comparable sociolinguistic data—will sociolinguistics be in a position to realize the robust contribution to the understanding of human language that its practitioners aim for.

In the decade ahead, Oxford Studies in Sociolinguistics will provide a forum for the most innovative and important studies of language in context. In so doing, it hopes to influence the agenda for linguistic research in the next century and to provide an array of insightful and provocative analyses to help launch that agenda.

Edward Finegan

Auld Ayr wham ne'er a town surpasses
For honest men and bonny lasses.

Tam O'Shanter

Preface

The work presented in this volume had its origin back in 1964–65 when I was a graduate student at University College, North Wales. During that year I came across William Labov's (1963) article "The Social Motivation of a Sound Change." This was the second-most-exciting experience I had had in the study of language (the first was reading Benjamin Lee Whorf). So I naturally made a point of attending Labov's seminar at the 1966 Linguistic Institute at the University of California, Los Angeles. I also made great efforts to digest and absorb the material contained in his *Social Stratification of English in New York City* (1966a). The pursuit of a Ph.D. combined with full-time teaching prevented me from undertaking any empirical research in the field until 1973. In that year a one-semester sabbatical leave combined with a grant from the Social Science Research Council enabled me to carry out a survey in Glasgow (Macaulay and Trevelyan 1973) that resulted in the publication of my *Language, Social Class, and Education: A Glasgow Study* (1977) and several articles (Macaulay 1975a, 1975b, 1976a, 1976b, 1978a). In 1978 and 1979 I conducted interviews in three other locations as part of a proposed comparative study of urban speech in Scotland. The three locations were Aberdeen, Ayr, and Dundee. Edinburgh was not included because a survey sponsored by the Social Science Research Council was already in progress (Romaine 1978). The aim of the survey was to investigate the extent to which similar patterns of socially stratified speech could be identified in each place and also to find out whether there were socially significant features of speech that were restricted to a single location. Administrative responsibilities at Pitzer College over the next six years prevented the completion of the interview schedule as planned, but I was already having serious reservations about the feasibility of the project.

The Glasgow study was modeled on Labov's work, though no attempt was made to replicate the scope and depth of that work. Partly, this was because of the time constraint. The whole project, including the writing of the final report, had to be completed in seven months. The other reason was that the linguistic investigation was only one part of the project. The rest consisted of interviews with teachers and employers. It was undoubtedly an overambitious project and resulted in several compromises that weakened the final product. The model for the linguistic part, however, was Labov's work in New York.

The Glasgow study produced a series of taped interviews with the adult respondents that, with one exception, could be considered comparable. The exception was a working-class respondent whose responses were fuller and more explicit, with the result that his interview lasted more than twice as long as any other. With this exception, the Glasgow interviews lasted approximately the same time and produced comparable amounts of recorded speech. These interviews provided the basis for an analysis of five phonological variables that showed various patterns of social stratification and age and gender differences. No attempt was made to extend the analysis to syntax or discourse features; but it would have been difficult anyway, given the limited nature of the interview data. Twenty-two preliminary interviews were deliberately excluded from the analysis because they had been collected by different methods and were not part of the quota sample. Even at the time, I felt that this was a serious loss, since they contained some good samples of Glasgow speech; but I believed that the design of the study required their exclusion. Fortunately, the tapes remain; and I hope one day to return to them.

In conducting the later interviews in Aberdeen, Ayr, and Dundee, I found greater variability in the interviews. Instead of lasting 30 to 40 minutes, some of the interviews took much longer, up to two-and-a-half hours. It was not only the quantity that varied but also the quality. There were many more narratives and extended responses to questions. The differences in quantity and quality from the majority of the Glasgow interviews are probably due mainly to two factors. The first is an improvement in interviewing technique. On listening to my own performance during the Glasgow interviews (and the process of analysis requires that one listen to the same passage many times), I was too often made aware of the places where I had missed an opportunity to encourage a respondent to develop a point; even worse, there were many occasions on which I had actually cut off a voluntary remark in my eagerness to ask the next question. In fact, my behavior suggests that I was concerned about the cost of recording tape and anxious that none of it should be wasted on silence. In the later interviews I was more patient, willing to let the respondent feel the responsibility of filling in the silence.

The second factor that affected the quality of the interviews was the method of selecting respondents. In the Glasgow survey all the adults were parents of children who had been interviewed, and they were contacted formally by letter with an official letterhead explaining in rather general terms the purpose of the survey. All the respondents were therefore aware that this was in some sense an "official" survey; and although I endeavored to put them at their ease as much as possible, their responses were doubtless affected by their perception of the nature of the speech event. They tended to reply rather narrowly to the questions and apparently did not feel free to digress as much as I would have liked. In the later set of interviews, many of the later respondents were identified through network contacts with the benefits of association that Lesley Milroy (1980) has demonstrated. Moreover, the purpose was more vaguely described as a book I was working on. This I am sure helped to create more of a feeling that the respondents were helping me directly rather than being an impersonal part of an official survey.

Whatever the reasons, I was fairly pleased with the quantity and quality of the speech contained in the later interviews; but as there was considerable variability between them, I was still concerned about their comparability. At the same time, the richness of the data provided an opportunity for a different kind of analysis, one that was not restricted to phonological or morphological variables. The present work might best be labeled a study in *discourse micro-sociolinguistics*. It deals with 12 of the interviews collected in Ayr in 1978–79. The interviews have been transcribed in their entirety and coded in terms of major syntactic units such as independent, coordinate, and subordinate clauses; infinitives; gerunds; quoted direct speech; and so on. This procedure provides a detailed way of charting the dynamics of the interviews, showing the effect of factors such as topic, genre, and the interview situation on the language recorded. The resulting profile of the speech event is independent of any subjective estimate of style such as "casual" or "spontaneous" and thus can be used as an objective correlative of any impressionistic views of the character of the interaction. Rather than assuming that the interviews are equivalent, the procedure provides a way to examine the significance of individual differences. Since the sample of speakers is socially diverse and there is considerable variation in the use of certain phonological, morphological, syntactic, and discourse features, it is possible to show which features are sensitive to social variation without the need to claim that the sample itself is representative of the whole population. The aim is to show the value of an approach that focuses on all the data collected rather than on variables taken out of context.

The last account of Ayr speech (Wilson 1923) is more than 60 years old, and the evidence was collected by very different methods. Nevertheless, as will become apparent, there has been considerable stability in the language over that period of time. I hope that the present description will, among other things, provide evidence of the speech of Ayr in the late 1970s that will be of similar value to future students of linguistic variation and linguistic change.

Any work that has been gestating for such a long period owes a great deal to other people. For their comments, advice, encouragement, and (not least) example, I am grateful to Jack Aitken, Donald Brenneis, Wallace Chafe, Alessandro Duranti, Charles Goodwin, John Harris, John Kirk, William Labov, Derrick McClure, Jim Miller, Suzanne Romaine, and Sandra Thompson. Special thanks for their comments regarding the final version are due to Douglas Biber, Nancy Dorian, and most of all Edward Finegan. Parts of this material have appears in a variety of published papers (Macaulay 1984, 1987b, 1988a, 1988b, 1988c, 1989a, 1991). The fieldwork was carried out with the assistance of a summer fellowship from the National Endowment for the Humanities and grants from the Research and Awards Committee of Pitzer College, to both of which I hereby express my gratitude. Finally, my thanks to Stanley George and Michael K. Lane of Oxford University Press for much needed help in preparing the version for the printer.

Claremont, California   R.K.S.M.
September 1990

Contents

Contents

LOCATING DIALECT IN DISCOURSE

I

The Sociolinguistic Study
of Dialect

Just because something is common does not mean that it is easy to understand. The common cold remains a costly and tiresome affliction despite the efforts of researchers to find a cure, while vaccines and antidotes have been developed for much less frequent diseases. The frequency with which human beings use language has not led to agreement on how they do it or even what it is that they do. In a single day a speaker may produce twenty thousand utterances with over 100,000 words; but unless this flow of speech is recorded somewhere in memory, on paper, or by some instrumental means, it will disappear without trace. Until recently, speech could be preserved only by painstaking methods that limited the kind of language that could be recorded and the circumstances in which it could be transcribed. Thus, it was not surprising that Bloomfield complained that "Saussure's *la parole* lies beyond the power of our science" (1927, 444). The invention of the portable tape recorder changed that.

Sociolinguistics, as widely understood nowadays, became possible with the invention of the portable tape recorder. The possibility of rehearing again and again examples of impromptu speech opened up opportunities for analysis of a kind that would otherwise have been inconceivable. Prior to the availability of substantial recorded examples of continuous speech, only relatively salient items could be observed on a wide scale. Labov was apparently the first linguist to realize the significance of this development. First in his study on Martha's Vineyard (1963) and later in New York (1966a), Labov demonstrated how quantification could be used to analyze the use of linguistic variables in a way that would allow for comparison between individuals and groups of individuals and also correlation with extralinguistic variables. The use of quantified index-

3

es to classify speakers distinguishes this approach from that of the ethnography of speaking (e.g., Gumperz 1971, 1982; Hymes 1974).

The existence of the tape recorder, however, did not solve the problem of what examples of speech to record. The problem is, How do we decide out of "the universe of potential observations" (Coombs 1964, 4) what to record? As Labov observed in his pioneering study of New York City, "bulk alone is no measure of adequacy" (1966a, 155). There has to be some process of selection. Labov's approach was to identify a random sample of the population to be interviewed in a systematic way, and his example was followed in a number of subsequent studies (e.g., Shuy, Wolfram, and Riley 1968; Trudgill 1974). In my own study in Glasgow (Macaulay 1977), mainly in order to save time, I used a judgment sample, in which respondents were chosen to meet certain criteria: occupational status, gender, age, and religion. In New York Labov was able to make use of a sample based on an earlier sociological survey. Perhaps because subsequent surveys have not had this luxury, the emphasis on random sampling has receded (L. Milroy 1987, 38), even in Labov's later work (1972). Romaine (1980a, 170) has gone further and questioned the adequacy of any random sampling method in sociolinguistic research.

The assumption behind random-sampling procedures is that it is possible to obtain comparable samples of speech from each of the respondents. One way to do this is to structure the interview so that the respondents are constrained in their responses to a similar pattern of speech. This is the method followed in traditional dialect questionnaires such as those used in the Survey of English Dialects (Orton 1962) and the Linguistic Survey of Scotland (Mather and Speitel 1975–86), where citation forms are elicited individually. This provides comparability at the cost of continuous speech. To obtain samples of the latter, Labov developed a strategy for the "sociolinguistic interview" (Labov 1981) designed to combine spontaneity with more formal styles in a manner that allows comparability. However, the equivalence of the interviews collected by such a method has generally been taken for granted; and usually, no account is given of the quality of the interviews. It is assumed that each speech sample is equivalent even though it is obvious that an interview with a voluble, willing respondent produces not only more but also qualitatively different speech from that produced by an interview with a reluctant, nervous, or tongue-tied respondent. This problem has not received much attention except from those (e.g., Wolfson 1976; Milroy and Milroy 1977) who have adversely criticized the quality of speech obtained in interviews. However, if the speech samples are not strictly comparable, the value of random sampling is lost.

The question of the comparability of the interviews has probably not arisen because the interviews as a whole have not been used as data. It is somewhat paradoxical that most of the speech collected in sociolinguistic surveys remains unanalyzed. Most investigators have followed Labov's lead in concentrating on a few variables. In this approach a certain number of tokens are extracted from the interview and coded. The analysis then deals with these tokens and the remainder of the interview is ignored. The variables analyzed are usually phonological, since such items "are high in *frequency,* have a certain *immunity from conscious suppression,* are integral *units of larger structures,* and may be easily *quantified on a linear scale*" (Labov 1966a, 49, emphasis original).[1] The concentration on such variables has, however, an influence on the kind of questions that are asked.

Labov's motivation has always been clear: his primary interest lies in the attempt to understand linguistic change (1966b, 102), and he specifically rejected the goal of linguistic description (1966a, v). The sociolinguists who have adopted Labov's methodology have tended to share his interest in change rather than description. However, lack of change may also be deserving of study (Macaulay 1988a). Moreover, not all features of language change at the same rate. The most salient changes are in vocabulary, but this is the aspect of linguistic change least studied by sociolinguists (Macafee 1988 is an exception). Instead, the analysis of the speech recorded in sociolinguistic interviews has largely been concentrated on tokens of phonological variables extracted from the text of the interview and considered only in the context of preceding or following phonetic segments.

The focus on linguistic change in sociolinguistic research has also concentrated attention on variables that show stylistic variation (*markers* in Labov's terminology). Labov has argued that the contrast in the use of a variable between monitored and unmonitored speech is one of the best indications of a linguistic change in progress. He has investigated these differences by examining speech in different contexts, from the informal to the formal. Unfortunately, style has proved to be a more complex phenomenon than Labov's early operational model (1966a) suggested. Bell (1984), Coupland (1988), Hymes (1972), Irvine (1979), Levinson (1988), and Traugott and Romaine (1985), among others, have pointed out the complexity of the situation. Moreover, attempts to control the stylistic context run into the basic problem that "the context of situation" (Malinowski 1935) is not only a physical reality but also a mental one. As Glassie has observed: "What matters is not what chances to surround performance in the world, but what effectively surrounds perfor-

[1]Notable exceptions are Lindenfeld 1969, Feagin 1979, and Broeck 1977.

mance in the mind and influences the creation of texts. That is context"
(1982, 521). It is important to realize that this mental context is created by the
participants in the course of the interaction (cf. Craen 1984).

Meanwhile, others (e.g., Chafe 1980; Du Bois 1987; Givón 1979a, 1979b;
Hopper 1979; Thompson 1983) who would not necessarily label themselves
sociolinguists have been using quantitative methods to look at the use of
language in a discourse context. In part, this is probably because of the
increasing interest in pragmatics and speech acts (Grice 1975; Levinson 1983;
Searle 1969) and the need to look beyond intrasentential relations. In particular, a concern with the given/new, topic/comment, theme/rheme distinctions,
deixis and anaphora, and iconicity in grammar has blurred the distinction between "linguistics" and "discourse analysis" (Brown and Yule 1983; Givón
1983). Other scholars interested in possible differences between written and
spoken discourse have also used quantitative methods (e.g., Biber 1986). In
general, however, in such studies there has been little concern with the type of
methodological data collection questions that L. Milroy (1987) raises. On the
contrary, it sometimes seems as if the choice of texts has been largely moti-
vated by convenience; alternatively, the language samples have been collected
in an experimental setting (e.g., the "Pear stories" in Chafe 1980). The
quantitative analysis in discourse studies has generally been applied to the
textual distribution of variables rather than to the characterization of speakers.
In such work there has also been little attempt to identify extralinguistic
variables that might affect the use of language. Even those who have used
speech samples collected as part of sociolinguistic surveys (e.g., Bower 1983;
Schiffrin 1987) have not related their observations on discourse features to
other aspects of language or to extralinguistic variables. Sometimes quantitative methods have been used to characterize genres (Biber 1986; 1988),
spoken versus written language (Tannen 1982), formal versus informal discourse (Thompson 1984), topic continuity (Givón 1983), and grammatical relations (Du Bois 1987).

There has thus arisen a paradoxical situation. Sociolinguists who are interested in the use of language in its social context have tended to ignore the
wider linguistic context in which their linguistic variables are embedded;
discourse analysts and others who pay attention to the larger linguistic context
have tended to ignore the social context in which the language is located. It is
difficult to believe that this division of labor is beneficial.

The present work is intended to illustrate how the totality of speech recorded in an interview can be used to provide a more complete description of
the use of language by members of the speech community. It differs in many
respects from previous sociolinguistic studies. It is not based on any systematic sampling procedure; rather, individuals were reached through network con-

tacts (cf. L. Milroy 1980). The interviews are not structured in any consistent pattern and differ greatly in length and variety. The major extralinguistic variable is social class; but the sample is unbalanced for gender (three women and nine men) and age (a range of 35- to 78-years-old). On objective grounds, it would be hard to claim that this is a representative sample of the community. Subjectively, as someone who was born and grew up only 20 miles from Ayr, I am confident that the interviews as a whole cover a wide stretch of the spectrum of linguistic variation in the community for this age group.

The analysis shows that there are considerable differences in the language used by middle-class and lower-class speakers. Some of these differences are fairly salient and "robust," for example, phonological and morphological features. Other features, such as the use of various clause types and discourse markers, are much more "fragile" indicators of social class membership, since their frequency depends more on topic and genre. Moreover, there is a range of variation within each social class group, as was found also in Glasgow (Macaulay 1978a).

An equally important (if not more important) aspect, however, is that detailed analysis shows the interviews to be self-validating in terms of the quality of speech recorded. Wolfson has argued that "the method of collecting sociolinguistic data by means of formal, scheduled interviews is at best inadequate and at worst counterproductive" (1976, 202). She applies these strictures even to the so-called spontaneous interview: "The so-called spontaneous interview, however, *is not a speech event*. It goes by no name which would be recognizable to members of the speech community and it has no rules of speaking to guide the subject or the interviewer. As anyone who has ever done this kind of fieldwork can testify, it can be an exceedingly uncomfortable thing to do" (p. 195, emphasis original). Of course, if the interviewer is uncomfortable, it is highly unlikely that the interviewee will be relaxed; but if the interviewer is at ease, there is a reasonable chance that the interviewee will also feel comfortable. The interviews on which the present work is based refute Wolfson's claim that the speakers have "no rules of speaking" to guide them. On the contrary, as will be shown in the chapters that follow, the speakers are making effective use of the discourse features and rhetorical devices of everyday conversation.

It is important to emphasize that although I identified the speakers through second-order network contacts (Boissevain 1974), I had not known any of them before. I scheduled the interviews and received their permission to tape record the events. There was no attempt to disguise the fact that this was an "interview," although I tried to make my questions and comments as conversational as possible. I believe that the interviews were on the whole successful because I managed to display genuine interest in what the speakers had to say, and the speakers noticed this. The less successful interviews were those in which the speakers could not believe that I was really interested in them or their experiences.

The quality of speech recorded in the Ayr interviews should help to reassure those who wish to carry out sociolinguistic surveys. Although group interviews are a valuable source of information, it is hard to arrange such events without creating an even more "unnatural" occasion than dyadic interviews. There may also be problems with voice identification if a single microphone is used. The kind of arrangements that Labov and his colleagues (1968) developed to record group sessions with the Thunderbirds and the Jets (using individual microphones) would be difficult to carry out on a large scale, particularly with adults. The dyadic interview is the easiest to arrange and to conduct. If other members of the family are present, it may contribute to the flow of speech; but (as will be shown in a later chapter) it does not always help.

The aim of the investigator should be to record an adequate amount of impromptu speech. If this means discarding certain interviews afterwards as unsuccessful, nothing is lost except time and effort. If comparisons are to be made between individuals and between groups, the samples of speech must be comparable to a certain extent. Although some of the interviews in the Ayr study are more than twice as long as others, they all contain enough examples of continuous speech to make comparison meaningful. This does not mean that they should be treated as equivalent—clearly, they are not equivalent. However, dealing with the total interview as a speech event allows factors such as difference of length and presence or absence of narratives to be taken into consideration.

The chapters that follow describe the study (Chapter 2); the method of analysis (Chapter 3); phonological, morphological, and syntactic features (Chapters 4–7); lexical differences (Chapter 8); and various discourse features (Chapters 9–11). Chapter 12 describes and evaluates the individual interviews, Chapter 13 summarizes the findings for the dialect of Ayr, and the final chapter draws some conclusions for sociolinguistic methodology.

To sum up, the aims of the present volume include the following:

1. To demonstrate that tape-recorded individual interviews with strangers can provide a wide range of linguistic information for comparative analysis
2. To show how the speakers make effective use of "the rules of speaking" in such speech events
3. To illustrate the use of quantitative methods to examine phonological,

morphological, syntactic, lexical, and discourse features in the context of the total interview

4. To identify the linguistic features that display variation according to social class
5. To provide a description of Ayr speech that will serve for comparative purposes with other dialects or future studies

The findings are obviously limited by the size and nature of the sample (partly the result of circumstances unforeseen at the time the original study was planned); but I hope that the illustration of the analysis will encourage other researchers to collect and analyze interviews of this type in a wide variety of situations. One of the problems I found was that there was very little comparable information available. I believe that we would understand much more about the nature and extent of linguistic variation if investigators would pay more attention to sociolinguistic descriptive studies.

2

The Ayr Study

The Study

Ayr was founded as a royal burgh in 1205. There was a sheriffdom in Ayr by 1207 and a Dominican friary there by the middle of the thirteenth century. During the Middle Ages Ayr was the chief Scottish port on the West coast. By the sixteenth century Ayr was a major center of commerce and a seaport with a thriving trade in textiles, leather products, and metalwork; but "the basis of Ayr's mercantile prosperity lay in her considerable trade with France, being the principal western seaport for the importation of wines, salt, and luxuries, exporting herrings, textiles and whatever else could find a market in France" (Strawhorn 1975, 56–57). After the deepening of the Clyde led to the development of Glasgow as a port, Ayr's importance in overseas trade declined; but Ayr remained the administrative center for a county that found a new source of prosperity in coal mining and dairy farming. Although Ayr increased its population from 29,000 in 1901 to 48,000 in 1971 (p. 194), at least some of this expansion came from incorporating the neighboring villages of Alloway and Whitletts; and the great majority of its inhabitants are Ayrshire-born. Ayr is perhaps best known to the rest of Scotland and the world beyond as the birthplace of the poet Robert Burns.

The Sample

The original plan for the comparative study of urban speech in Scotland called for a gender-balanced, socially stratified, sample of 10-year-olds, 15-year-olds, and adults to allow for comparison with the Glasgow study (Macaulay 1977). As pointed out in the Preface, circumstances prevented the completion of the project as planned. A total of 62 interviews were carried out in Ayr, including 18 adults, 21 15-year-olds, and 23 10-year-olds. The children were identified through the school system and interviewed on the school premises. The adults were, in most cases, contacted through network connections, which as Lesley Milroy (1980) has shown, is likely to provide a better environment for recording spontaneous speech. The analysis in the present work deals only with 12 of the adult interviews obtained through network contacts. Six of the 12 adults can be characterized as middle-class and 6 as lower-class. In order to protect their anonymity, they will be described in only very general terms; and the names are fictitious.

All 12 were interviewed in relatively similar conditions (in their homes, with one exception), and the interviews followed similar paths in their treatment of topics. An additional six adult interviews were much shorter and more idiosyncratic in nature. There was nothing in these six interviews that differed radically from the interviews analyzed, but there was simply insufficient speech to justify including them in the overall comparison. The interviews with the 10-year-olds and the 15-year-olds were generally unsatisfactory. For some reason, I was less successful in most cases than I had been with the schoolchildren I interviewed in Glasgow. It is possible that the Ayr children were less sophisticated or street-smart than their Glaswegian equivalents. The tapes remain for future analysis, but a preliminary examination does not suggest that they will prove very illuminating.

Middle-Class Respondents[1]

Wallace Gibson (WG). Age 55. Lived all his life in Ayr. Active in cultural affairs. Father was in the insurance business.

Andrew MacDougall (AM). Age 35. A scientist with a Ph.D., engaged in postgraduate study. Had spent some time away from Ayr. Father had been a greenkeeper, then a janitor.

James MacGregor (JM). Age 63. A schoolteacher. Lived all his life in Ayr except for army service during the war. Father had been a small businessman. Interviewed along with Nan Menzies, whom he had known since their school

---

[1] All the names are fictitious and the descriptions are deliberately vague. Although I believe there is nothing offensive in the accounts of their speech, I did not want the speakers to be embarrassed in any way by the comments on their speech.

days. Mrs. MacGregor was present during the interview and contributed a few remarks, occasionally prompting her husband.

Nan Menzies (NM). Age 62. Widow of a schoolteacher. Father had been a building contractor. Had lived most of her life in Ayr but had gone away to college in Edinburgh and had spent about a year in England before World War II. Interviewed with James MacGregor.

Duncan Nicoll (DN). Age 67. Left school at 14. Had held a variety of white-collar jobs including tailoring, insurance, and the hotel business. Father had been a tailor, later a hotel keeper. Mrs. Nicoll was present for a brief period but did not speak much.

Ian Muir (IM). Age 59. A successful businessman. Father had been a lawyer. Attended a private boarding school away from Ayr. Apart from that and army service during World War II, had lived in or near Ayr all his life.

Lower-Class Respondents

Hugh Gemmill (HG). Age 78. A farm laborer, later a factory worker. Left school at 13. Lived all his life in and around Ayr. His father had been a shepherd.

Ella Laidlaw (EL). Age 69. Married twice to men who held manual jobs at a factory. Left school at 16. Lived all her life in Ayr. Father had been a coal-man. Her second husband was present during the interview and contributed occasional remarks and narratives.

Willie Lang (WL). Age 56. Had been a coal miner much of his working life. At the time of the interview was working as a caretaker. Left school at 14. His father had been a coal miner.

Willie Rae (WR). Age 54. A coal miner. Father had been a coal miner but died when Willie was eight. Brought up by his grandfather, who had also been a coal miner. Left school at 14. Lived all his life in or near Ayr.

Mary Ritchie (MR). Age 70. Had worked in a carpet factory and as a house-cleaner. Widow of a woodcutter. Her father had been a coal miner. Left school when she was 14. Had lived all her life in Ayr.

Andrew Sinclair (AS). Age 58. Had been a coal miner. Now working as a lorry driver. Father had been a coal miner. Left school when he was 14. Had lived all his life in Ayr except for a period in the merchant navy during World War II.

Social Class

It is almost impossible to deal with social class in an uncontroversial manner; yet the persistence of social class differences in Britain is clear, and social class is one of the extralinguistic factors that often shows a correlation with language variation. Moreover, the attention that the work of Bernstein (e.g., 1962, 1971) has received makes the question of social class differences in language a vitally important one. As will become apparent in some of the findings presented later, there is little support for Bernstein's position in the Ayr interviews.

A number of sociolinguistic studies (e.g., Labov 1966a; Trudgill 1974; Wolfram 1969) have based the assignment of individuals to a category on an index calculated on a variety of dimensions, such as occupation, income, education, housing, and so on. As I pointed out in an earlier paper (Macaulay 1976a), the use of multidimensional indexes to determine social class is problematic, as there is no way to rank the different factors on equal interval scales and consequently no way of identifying the amount of distortion caused by multiplying or adding together different factors. Such methods are necessary when the speakers are drawn randomly from the population, but in a judgment sample individuals are chosen on the basis of certain characteristics. In the Glasgow study (Macaulay 1977), the speakers were chosen to represent four occupational categories derived from the registrar-general's census classification. In the present study, the speakers were identified through network contacts in the community and assigned to one of two social classes: *lower* or *middle*.[1] The accuracy of the assignments is confirmed by the background information contained in the interviews themselves.

First of all, there is occupation. All the middle-class speakers, and none of the lower-class speakers, held (or had held) white-collar jobs. Moreover, with the exception of MacDougall, there had been no change of occupational status from the parents. As Sinclair observed, "Your father was a miner, you were a

---

[1]Even the choice of labels is controversial. I have chosen to use *lower-class* rather than the euphemism *working-class*, but it would have been possible to use numbers or *blue-collar/white-collar* or a variety of other labels. The choice of the labels *middle-class* and *lower-class* is not *intended* to have ideological implications but cannot avoid them.

miner, sort of style." In fact, that could be a reason for choosing a job. Ritchie's father had been a gardener; but when his sons were growing up, "he went to the pits . . . for the sake of getting them work." Rae, whose father had died when he was eight, explained how he was 20 before he got to work down the mine: "A lot got before that, because their fathers were living. Well, I had nae father, so I had to start at the bottom, ken." In contrast, Sinclair was only 14: "I went down the pit, and my father had a good name in the pit for his work and that he done, and through that I only worked in the pit bottom maybe for about three or four days, and I got a job—what they called going into the insides, that was away into the booels of the earth."

The middle-class speakers had followed their parents in similar, if not always identical, occupations. The exception is MacDougall. His parents were lower-class, but he had achieved middle-class status through education. He commented on his awareness of the difference between his primary and secondary school experience:

> When I was at primary school, well I just played with the local children, and, you know, everyone that lived in the same neighborhood played together quite often at football or cricket or just anything at all. And when I went to Ayr Academy, then I was mixing with the people from other parts of the town, and I found that I didn't really see a lot of them after four o'clock, because they all went back to their own areas. . . . I think it was partly a geographical thing. The South side of Ayr tends to be much more upper-class than the North or East side.

Again, with the exception of MacDougall, the middle-class speakers, at the time they were interviewed, all lived in the southern part of the town. MacDougall lived in a council flat that he had taken over after his father died because at the time of the interview he had very little income, as a full-time student. The lower-class speakers, with the exception of Gemmill, lived in Whitletts, which had been a separate village until its incorporation into Ayr just before World War II. When Sinclair and Lang were growing up, there had been only three streets; and they were used for football and other games:

> Well the High Road played the Low Road, and the Low Road played the Main Road, et cetera and that. I mean there was always that competition between the roads more or less. (Sinclair)

The closeness of the village to the countryside meant that there was an opportunity for the children to help and work on the farms:

> And it was great. You know, you'd sit up in the cairt, driving the horse, going to the station with the tatties, et cetera, and then sitting on the tap of the long cairt when they were cairting in the hervest, and aw these different things. (Sinclair)

The countryside also provided opportunities for games, as Ritchie remembered:

> We had wer girds [metal hoops] too. We all had wer girds, and the boys cried the
> girls a race, and we left that gate. And one lot went up the High Road and right
> roond about and doon by the graveyard, and the other lot went up along the
> Straiton Road and up round—to see who could come back to that gate first.

Games of this kind and football were frequently mentioned by the lower-class
speakers. The middle-class speakers, on the other hand, made frequent reference to cricket, golf, rugby, and tennis—games that do not crop up in the
lower-class interviews.

This is obviously related to schooling as well as to the cost of equipment
and facilities. All the middle-class speakers had enjoyed their schooling and,
with the exception of Nicoll, had stayed on beyond the minimum school-
leaving age. Four had gone on to forms of higher education, and Muir had
chosen to go into business after the war rather than continue his legal apprenticeship. Only Nicoll had left at 14, in order to enter his father's trade as a
tailor. In contrast, all the lower-class speakers had left school as soon as they
could. As Mary Ritchie put it, "I wasnae a lover of the school"; and that was
generally true of the others. Only Laidlaw had done well at school, and her
mother had kept her there until she was 16; but then she insisted on leaving: "I
just wasnae interested. I was more interested in devilment and fun."

The middle-class speakers remembered parties, family activities, and holidays [vacations]:

> Games like Spin the Plate and Forfeits and all sorts of games like
> that . . . Postman's Knock and similar little things. . . . We played charades. . . . We used to have the Scottish students songbook; and one member of
> the family played, and everybody gathered round and sang all these happy
> songs. (Menzies)

The lower-class speakers did not go for vacations when they were young;
and their memories of home were more of orange boxes for furniture, "the
boyne" (a tub) for washing, outside toilets, and chalk on the floors. When
Gemmill collected his retirement pay, he "put carpets on aw the flairs in the
hoose wi the money. That was handy."

There was also a clear difference between the two groups in their attitude
toward the past. The middle-class speakers tended to look back nostalgically,
particularly as regards the kind of service that was no longer provided:

> People used to come into the shop and say: "Oh it's so nice here. It's not like
> this in England, you know." And the inference being that they were well
> attended to. But not, by God, now. Sometimes it's appalling. And the trouble is,
> you see, it won't get any better until there's a change of attitude. (Muir)

In contrast, Gemmill, who had received five-pound wages for six months as a farm laborer, had a very different view of the past:

> They talk aboot the good old days. There nothing good aboot them. . . . [RM: Is there anything you think was better in the old days?] Not that I ken of, naw. The fact is you've got money noo that you hadnae, and money's—it's maybe the root of all evil, but it's a necessity tae at the same time.

Thus on the basis of family background, education, occupation, residence, interests, and attitudes, there is a clear difference between the two groups of speakers. There is no need to construct any index to justify the assignment of any speaker to a particular category. In Broeck's (1984, 51) terms it is a "polarized" sample. The fact that (as will become clear in later chapters) there is considerable variation within the speech of each group need cause no surprise (Macaulay 1978a; Klein 1988) and does not call into question the assignment of any speaker to one or the other class.

Gender

The absence of reference to gender in the present study should not be taken as indicating a lack of interest in this factor or a belief in its irrelevance. Rather, it is a result of the unbalanced gender distribution of the sample. With only three women, two of whom (Menzies and Ritchie) produced relatively small samples of speech, it would be difficult to make a meaningful comparison with the male speakers.

The Data

As will become apparent, the interviews differ in their character. Some respondents show greater initiative in taking control of the speech event, while others are more dependent on questions and directions from the interviewer. Since the interviewer's contributions are also coded, it is possible to tabulate their frequency as well. For simplicity in this aspect, the interviewer's remarks were sometimes abbreviated in the transcription if this did not appear to distort the interaction. The interviewer's back channel responses (e.g., *yes, mhm,* etc.) were also omitted from the transcription, although such features are worthy of separate study.

The first significant factor is the length of the interview. The interviews ranged in duration from 40 minutes to two-and-a-half hours. Obviously, the longer the interview, the more opportunity there is for the respondent to say

Table 2.1. Interview Length

Respondent	Number of Words	Number of Syntactic Units
Lower-class		
Ritchie	4,373	745
Rae	5,030	744
Gemmill	9,761	1,697
Laidlaw	13,189	2,489
Lang	16,255	2,319
Sinclair	21,163	3,208
All LC	69,771	11,202
Middle-class		
Muir	4,573	813
Menzies	5,110	839
MacGregor	7,673	1,112
Gibson	8,510	1,519
MacDougall	9,764	1,477
Nicoll	15,268	2,679
All MC	50,898	8,439

something and the greater the possibility of different genres occurring. Table 2–1 gives the number of words and the number of syntactic units in each interview. It can be seen from Table 2–1 that there is a wide range; and while the differences are not solely a function of social class, the lower-class interviews on average are somewhat longer and (as will be shown later) contain more narratives and a larger number of topics. Differences in the length of the interviews, however, are not a function of the number of questions asked. On average, it turns out that I asked about 50 questions, but in two of the shortest interviews I asked about 50% more. The number of questions and the proportion of syntactic units and words in response are shown in Table 2–2. The inverse relationship between the number of questions and the length of the interview is hardly surprising. Where the interview flows smoothly, there is no need for the interviewer to intervene. It is when the respondent stops speaking that the need to prompt arises. As will be shown in a later chapter, differences in the kind of interaction in the interview affect the kind of language used.

This is a very crude measure of interaction. Following Albris and his colleagues (1988), I also looked at the number of topics that I had brought up and the number initiated by the respondent. Like many others in the analysis

Table 2.2. Average Length of Responses to Interviewer's Questions

Respondent	Number of Questions	Avg. Number of Syntactic Units per Question	Avg. Number of Words per Question
Lower-class			
Ritchie	78	10	56
Rae	76	10	66
Gemmill	53	32	184
Lang	49	47	332
Sinclair	47	68	450
Laidlaw	24	104	549
Middle-class			
Menzies	61	14	84
Muir	49	17	93
MacGregor	45	25	171
Gibson	46	33	185
MacDougall	51	29	191
Nicoll	11	244	1,388

of discourse, the term *topic* is differently defined by different analysts (e.g., Brown and Yule 1983; Givón 1983; Grimes 1975; Hurtig 1977; Keenan and Schieffelin 1976; Venneman 1975). As a starting point, I followed Van Dijk's notion of an *episode:* "a sequence of sentences (or of propositions expressed by such sentences) . . . with marked beginning and end and some conceptual unity" (1982, 179). As will be shown in Chapter 12, the Ayr speakers used a number of specific devices in introducing new topics. These topics are also *speaker's topics* (Brown and Yule 1983, 87–94), which they describe "in terms of a process in which each participant expresses a personal topic within the general topic framework of the conversation as a whole" (p. 88). The proportion of topics initiated by the speakers is shown in Table 2–3.

Table 2–3 shows that overall, the speakers were responsible for raising half of the topics and that more than half of the total time on the tapes was devoted to the discussion of topics initiated by those who were being interviewed. This makes the term *respondent* somewhat ironical. For the most part, the individuals whose speech was recorded were not passive subjects waiting for the next item on a questionnaire but active participants in a conversation that had its own momentum. The exceptions were Ritchie, Rae, and Muir, who were not the most forthcoming of respondents. Essentially, they responded to questions. Though they all gave extended replies on occasion, they seldom took charge of the interaction to launch into a topic that I had not raised or initiate

Table 2.3. Proportion of Topics Initiated by Respondents

Respondent	Topics Initiated by Respondent	Proportion of Total Interview Initiated by Respondent
Lower-class	%	%
Ritchie	13 (4/30)	4
Rae	13 (7/55)	15
Gemmill	23 (23/42)	67
Sinclair	56 (43/77)	50
Lang	77 (72/93)	82
Laidlaw	85 (52/61)	68
All LC	56 (201/358)	59
Middle-class		
Muir	12 (4/34)	12
MacDougall	26 (12/46)	20
Menzies/MacGregor	35 (36/104)	28
Gibson	44 (30/68)	44
Nicoll	78 (63/81)	82
All MC	44 (145/333)	44
All speakers	50 (346/691)	53

Note: Since Menzies and MacGregor were interviewed together, their joint scores indicate the proportion of topics that were not initiated by the interviewer.

other exchanges. Nicoll is at the other extreme. Without prompting he told me his life history in a sequence of sketches, and it is literally in the middle of the interview that I managed to ask my first question. Generally, there are more questions in the first part of the interview, with also a tendency for their frequency to increase again toward the end of the interview. On several occasions, Nicoll refers to himself as "the chatty type" or "the gossipy type"; and he is clearly proud of his ability to talk to a wide range of people, including, obviously, his present interviewer. Laidlaw is almost as prolix. As mentioned earlier, her husband was present during most of the interview and participated from time to time, so that the interview was more of a conversation. She also expressed surprise when I had to change the tape, as she had apparently been unaware of the recorder up to that point. However, her speech afterwards was no more inhibited. She also had the best sense of humor of the whole sample. The interview with Menzies and MacGregor should have had a similar atmosphere but neither respondent completely relaxed. There was a little interaction between the two; but for the most part they responded serially to questions. This may have been largely my fault for asking too-specific questions—though again, from time to time both gave extended responses. The interviews with Gemmill, Lang, and Sinclair were remarkably easy. All three took my questions as invitations to speak at length and needed little prompting. The interviews with MacDougall and Gibson were somewhat strained at the beginning. These were the two best-educated respondents and therefore ought to have felt most at ease with an academic interviewer, according to Wolfson (1976, 197); but this was not the case, although both relaxed considerably as the interview proceeded. These two interviews, and that with MacGregor, are characterized by frequent hesitations and filled pauses. A more detailed account of each interview will be provided in Chapter 12.

I have presented a general characterization of the interviews in preparation for the detailed examination of phonological, morphological, syntactic, and discourse features. Since a number of factors can influence the variation in these features, there is a risk in generalizing from either the group or the individual frequencies; but in many cases a clear pattern of social class variation will emerge. A major factor that affects the pattern of social class variation is the use of language in different discourse genres. Idiosyncratic factors affecting this variation will be discussed in Chapter 12, where I present a comprehensive account of each interview.

3

The Method of Analysis

The interviews with the 12 speakers listed in the previous chapter were transcribed in their totality in a modified orthography to show certain dialect pronunciation features. The interviews were also coded to show major syntactic features as shown in (1):

(1)   ca   coordinate clause introduced by *and*

        cao   coordinate clause introduced by *and* with no surface subject

        cav   coordinate clause introduced by *and* with no finite verb

        cb   coordinate clause introduced by *but*

        cbo   coordinate clause introduced by *but* with no surface subject

        cbv   coordinate clause introduced by *but* with no finite verb

        co   coordinate clause introduced by *or*

        coo   coordinate clause introduced by *or* with no surface subject

        cov   coordinate clause introduced by *or* with no finite verb

        cs   coordinate clause introduced by *so*

        cso   coordinate clause introduced by *so* with no surface subject

        csv   coordinate clause introduced by *so* with no finite verb

        ds   quoted direct speech

        fr   fragment in response to interviewer's question or comment

        ge   gerund

        im   imperative

        in   infinitive

        ma   main clause beginning with subject

        mc   main clause with *it*-clefting

        md   main clause with demonstrative focusing

        me   main clause with extraposition

mf	main clause with adverbial first
ml	main clause with left dislocation
mo	main clause without subject
mr	main clause with right dislocation
mt	main clause introduced by *there*
mv	main clause without a finite verb
my	main clause with Y-movement
nw	noun clause introduced by WH-form
pp	present participle
qu	interrogative clause
rn	nonrestrictive relative clause
ro	restrictive relative clause with Ø relative marker
rs	nonrestrictive sentence relative clause
rt	restrictive relative clause with *that*
rw	restrictive relative clause with WH–relative marker
sa	subordinate comparative clause introduced by *as*
sb	subordinate clause of reason
sc	subordinate clause of concession
sd	subordinate clause of comparison introduced by *than*
si	subordinate conditional clause
sm	subordinate clause of manner
sn	subordinate noun clause
sp	subordinate clause of place
sq	embedded question
sr	subordinate clause of result
ss	subordinate clause of purpose
st	subordinate clause of time
sw	subordinate clause of reason

Syntactic units were chosen as the basis of analysis to see whether there are social class differences in the use of syntactic structures. The choice of units depends on the aims of the analysis. Chafe (1986a) has used "idea units" and "intonation units," while Halliday (1967) has "information units." Although the status of syntactic units is perhaps less secure than it might seem (Thompson 1984), there may be some advantages to dealing with traditional categories. Since one of the aims of the present analysis is to provide numbers that can be used in comparative studies, it seems preferable to choose units that have the least subjective element in their identification. The syntactic analysis for each speaker is given in Appendix I.

The coding shown in (1) allows the sorting procedure in the word-process-ing program to group together all the clauses of a particular type for tabulation and further analysis. Another commercially available program produces concordances and also provides a word count, so that the average length of clause and the frequency of a particular structure can be easily calculated. The clauses are numbered sequentially, making it possible to observe the distribution of each structure throughout the interview. The search procedure of the word-processing program also makes it a simple matter to chart the occurrence and frequency of dialectally marked forms such as *ken* or *yin*.

It must be emphasized, however, that the process of coding in syntactic units is by no means as simple as it might appear. There are numerous occasions when more than one alternative is possible, and the choice is often arbitrary. Since I did all the coding myself and made every effort to be consistent, the arbitrary decisions probably do not affect comparisons within the sample; but comparisons with tabulations from other methods of scoring may not be valid. Essentially, the interviews are coded by surface structure syntactic units with one verb. The greatest problems arise with conjoined structures, particularly where a subordinate clause is followed by a clause introduced by *and, but,* or *or.* In some cases, it is clear that the structure is parallel to the previous subordinate clause; but many other cases are ambiguous. Where there is any doubt about the relationship, such structures are coded simply as coordinate clauses. Another problem is caused by hesitations, false starts, and repairs. Where the verb is repeated in more or less the same form, the sequence is treated as a single structure. Where the repair involves a different verb, the structures were coded as separate. Other problems include distinguishing between participles and gerunds and between modals and main verbs. In the latter case, an arbitrary decision was made to treat *going to* and *used to* as modals and all other examples of *to* + verb as infinitives. Finally, lexicalized expressions such as *I think* (when used parenthetically), *you see, I mean,* and *you know* were considered discourse features (see Chapter 10) and not counted as verbs for the purpose of identifying syntactic units.

The interviews are transcribed in a modified orthography that shows most but not all of the major phonological variables. There is no easy solution to the problem of transcription because a phonetic transcription provides too many irrelevant details. As Ochs has pointed out: "One of the important features of a transcript is that it should not have too much information. A transcript that is too detailed is difficult to follow and assess. A more useful transcript is a more selective one" (1979, 44). Moreover, as Haugen points out in connection with the specific question of representing a dialect:

> The guiding principle in any transcription must be that it should convey the information which its reader needs, no more and no less. Conveying more information than needed is to overwrite the dialect, and less than needed is to underwrite it. The writer must gauge his prospective audience's previous experi-

ence and temper his transcription accordingly. He must be able to judge which rules of reading his audience knows and can therefore assume without entering them in the transcription. Whatever decision he makes, it will therefore reflect a judgment concerning his readers. (1980, 276)

The problems of representing Scottish dialect are discussed elsewhere (Macaulay 1991). The transcription used in the present work follows the principles set out in that article:

(2)   I. A transcript should be appropriate for the specific purpose in which it is to be used.

  a. There is a difference between a transcript intended for the purposes of analysis and one to be used to illustrate certain points in an article to be read by people who have no access to the original.

  b. The representation of particular features may be clear to those who are familiar with the original and with the speech community in which the speech was recorded, but the same representation may be opaque to those unfamilar with either. In the latter case, some key to the transcription may be necessary.

  c. The inclusion of a representation of features that are irrelevant for the immediate purpose may have the effect of distracting the reader's attention from the essential features.

  II. The aim of any transcription is to make the reader's task as simple as possible.

  a. Any feature that complicates the reader's task constitutes "noise" (see Ic above) and should be avoided.

  b. The transcription should be designed with a specific type of reader in mind, not for universal purposes.

  III. It is unnecessary to indicate phonetic features which are predictable from general rules of the orthography.

  a. In English it is unnecessary to indicate such features as vowel reduction, consonant cluster simplification, and assimilation, which would be "normal" in reading aloud a text "in conversational style."

  b. Any "variant spellings" should be accompanied by a key indicating their phonetic quality and the social significance of the variants.

  IV. Indication of prosodic and paralinguistic features is best conveyed by a representation that is separate from the orthographic text.

    a. Where intonation is important for a particular point, it should be shown by a representation of the contour (e.g., as in Labov and Fanshell 1977).

    b. Discussion of prosodic features should include reference to the norms of interpretation in the speech community for such features (e.g., Local 1986).

    c. Indication of paralinguistic features should be limited to features relevant to the point under consideration.

V. The same passage might appear in several different transcriptions, depending upon the features under consideration.

    a. Any transcription can give only an indirect impression of a spoken text; it is misleading to suggest that a complicated transcription is "more accurate" than a simple one.

    b. The reader will bring certain skills to the task of interpreting an orthographic text, and the transcript should take advantage of these interpretative skills rather than disrupting them.

VI. The success of a transcription is not to be judged on how much the transcriber has managed to include but on how much the reader succeeds in getting out of it.

However, some features (such as the loss of voiced velar stops after homorganic nasals) are difficult to represent without recourse to phonetic symbols. The solution that has been adopted here is to use spellings taken from *The Concise Scots Dictionary* (Robinson 1985) as much as possible except the use of the apostrophe. For example, the vocalization of /1/ in *all* [ɔ:] is represented by *aw* rather than by *a'*. These spellings should be generally transparent even to those unfamiliar with Scottish dialects. In a few cases where no simple orthographic solution is available, a phonetic transcription is given in parentheses. No attempt is made to represent the normal processes of vowel reduction or consonant elision, since readers are accustomed to interpreting this in reading any texts. The major omission in representing the phonological variables is the glottal stop. Since this variable is discussed in detail in Chapter 4, individual representation of glottal stops would have been redundant.[1]

The coding and transcription are illustrated in (3) with an extract from the interview with Gemmill:

(3)   (Did you get any holidays [vacations]?)
    fr   411   oh aye aye but aye

[1]For consistency the spelling in the examples, except when indicating Scots pronunciation, follows the same convention as in the text, i.e., American. Obviously, such spellings are not intended to suggest American pronunciation.

ma	412	well the thing aboot the holidays is
si	413	as long as you werenae mairrit
ma	414	you were right enough for holidays [vacations]
me	415	it was the poor man
rt	416	that was married
ma	417	he got nae holidays [vacations]
nw	418	well what he got
ma	419	was this
ma	420	the feeing fair was aye on a Tuesday the second Tuesday in October and the second Tuesday in—eh April
ca	421	and
si	422	if you went to the fair
ca	423	you didnae get to Ayr Cattle Show
cb	424	but the single [sɪŋəl] man he—
me	425	it didnae maiter
sq	426	whether he was staying in the place
ro	427	he was in
sq	428	or shifting
ma	429	he got a week's holidays [vacation]
ca	430	and—eh—
nw	431	what a lot o them did
ca	432	they got a week's holidays
ca	433	and they didnae go to their place
cv	434	and the Tuesday after the fair—after the—
ma	435	well the twenty-eichth of Mey was the term
ma	436	well the following Tuesday efter that date was
nw	437	what they called "Duds Day"
ma	438	that's
st	439	when you went
in	440	to buy your claes for the next six month
in	441	to keep you
ca	442	and of coorse they got twaw or three days holiday again
ge	443	through daeing these things
ma	444	that's the kind
rt	445	that's oot for—
rt	446	that didnae want
in	447	to work really

The first column shows the coding for the syntactic unit. The second column contains the line numbering. Where the syntactic unit is too long to be accommodated on one line, as in line 420, the subsequent lines are not numbered. When a syntactic unit is separated by another syntactic unit, the second part is indented and numbered but the indented lines were deducted from the total count for the structure. Lines 422–24 illustrate the situation when another clause intervenes:

(4)   ca   422   and
         si   423   if you went to the fair
         ca   424   you didnae get to Ayr Cattle Show

The sequence "and . . . you didnae get to Ayr Cattle Show" is treated as a single coordinate clause introduced by *and*. Unfortunately, things are not always so simple. Line 435, "and the Tuesday after the fair—after the—," is treated as a coordinate clause, although line 437 could be considered to be its completion after the repair in line 436. Line 437 was coded as a separate main clause because of the discourse marker *well*, which does not usually occur after coordinating conjunctions. Problems of coding arose mainly after repairs, false starts, or repetitions.

The modified orthography shows certain features of Gemmill's speech:

(5)   412   *aboot* [əbut]
      413   *werenae mairrit* [wɛɾnë meɾët] (cf. *married* [mæɾëd] in
            416)
      417   *nae* [ne]
      426   *maiter* [meʔəɾ]
      436   *eichth of Mey* [exθ əv məɪ]
      437   *efter* [ɛftər]
      441   *claes* [klez]
      443   *coorse* [kurs]
      443   *twaw* [twɔ]
      444   *daeing* [deɪŋ]
      445   *oot* [uʔ]

There is no attempt in the orthographic transcription to show the absence of a velar stop in words such as *single* [sɪŋəl] (425) or the presence of glottal stops in *maiter* (426), *lot* (432), and so on, or precise vowel quality. Where such details are of interest, they can be indicated by phonetic transcriptions, as shown on line 425. Since the scope of the present work does not include the study of intonation, no attempt is made to represent prosodic features. Self-repairs are indicated by dashes (—) at the intonation break.

The coding thus provides a way of tracking the use of certain key features—phonological, morphological, and syntactic—throughout the inter-

---

[1]All the interviewer's remarks were coded separately as comments (2) and questions (2a) so that they could be tabulated.

view. In this way the effects of time, topic, and genre can be observed in a
way that avoids the need for such distinctions as Labov and the Milroys have
drawn between formal and casual speech or interview style and spontaneous
style. By definition, all the transcripts represent interview style, since the
respondents knew that they were being interviewed and recorded. The interesting point is to see what kind of speech has emerged from this exchange.
The analysis takes account of changes within the interview and also allows
comparison with the other interviews either on an individual or a group basis.
Ideally, this book should contain the transcripts in their entirety; but that
would have made it prohibitively long. Instead, I shall present an account of
the major dialect features, showing any variation within or between the two
social class groups. Aspects of language are treated roughly in order of their
salience as indicators of social class differences. Accordingly, the more
"robust" features are considered first, while those more affected by topic or
genre are dealt with later. As already mentioned, Chapter 12 will present an
analysis of the individual interviews, with extracts to show the interaction of
various features, and will attempt to isolate the idiosyncratic features of an
individual's speech from those that are characteristic of the speaker's group
membership.

A Note on Statistics

The first thing that will strike some readers is the total absence of any tests of
statistical significance in the tables. This is a deliberate omission. Despite
urging from a number of colleagues, I decided that I would be providing such
information insincerely, since I have become increasingly uncomfortable with
the use of statistics in this kind of linguistic investigation. It may be a
groundless fear but I am worried that relatively minor features may be given
greater importance than they deserve simply because they meet some criterion
for statistical significance.[1] The raw numbers for each variable are there for
those who wish to carry out such calculations, as has been done with some of
the results from the Glasgow study (Berdan 1978; Cichocki 1988). Berdan
and Cichocki, however, were using the material mainly to demonstrate methods of analysis rather than to draw conclusions from the particular data set.
My own view of the quantitative measures in the present work is that there are
too many uncontrolled factors for the results to be taken as anything more than
suggestive. Some readers may consider such a view to be inconsistent with the

---

[1] A cautionary example can be seen in the attempts to identify gender differences in language
development (Macaulay 1978b).

considerable amount of quantitative information provided in the following pages, but I believe that it is helpful to count instances and compare frequencies, since the results may focus attention on hitherto unnoticed aspects of language use. The numbers themselves, however, should not be taken too seriously. One day a historian of social science may examine the influence of statistical analyses on the way investigators have viewed their data. In the meantime, I prefer to side with Romaine (1982a) in thinking that attempts to make linguistics a quantitative science are premature.

4

Phonological Variables

Wilson (1923) based his account of Ayrshire pronunciation on five elderly "authorities" who lived in Tarbolton. Although Wilson does not discuss his methodology, he presumably collected his information by eliciting individual words and by listening to the conversation of his five authorities. Since he seldom gives variants, it can be assumed that the authorities generally agreed. Such consistency is common when citation forms are elicited and no attempt is made to press for alternatives. Similar consistency is found in the interviews of the Linguistic Survey of Scotland (Mather and Speitel 1975–86, vol. 3). Evidence based on tape-recorded interviews of connected speech is inevitably much messier. A variety of factors such as mood, rate of utterance, topic, and genre can affect the forms of speech. This point can be seen most clearly in the use of the vowels.

Wilson presents his analysis of Ayr pronunciation by showing how it differs from standard English, but he makes it clear that this is merely for convenience of exposition:

> In making the following comparison between Scotch as now spoken in Central Ayrshire and standard English, I do not mean to suggest that the one is derived from the other, but merely to point out the differences between the words as pronounced in the two forms of speech; which are really, as regards the words most commonly used, separate developments of the Old English group of dialects. Nor do I suggest that, in every case in which I compare a Scotch word with an English word similar in pronunciation and meaning, the two words have the same etymological origin. (1923, 15)

A similar caveat is appropriate for the present analysis. I will assume that when Menzies says *home* [hom] and Rae *hame* [hem] they are using forms

30

that are in some sense equivalent or in alternation; but no attempt will be made to justify this assumption or link them to a possible common ancestor in Old English *hām,* since that task requires a different kind of evidence from that to be found in the Ayr interviews. The assumption is necessary, particularly in dealing with the lower-class interviews; but it remains an assumption.

Consonants

Wilson provides a description that can be interpreted as the following inventory of consonants:

Stops: p t k, b d g
Nasals: m n ŋ
Affricates: tʃ dʒ
Fricatives: f v θ ð s z ʃ ʒ x
Glides: ʍ w j
Liquids: r l

In the Ayr interviews the consonant inventory remains the same, though the distribution of the sounds differs in some respects. Wilson does not give many phonetic details for the consonants. He does so for the sound /r/, which he says "is pronounced much more distinctly than in E[nglish], with a marked trill, due to the vibration of the tip of the tongue." Trilled [r] does not occur in the later interviews; only the alveolar tap [ɾ] and the postalveolar fricative [ɹ] are heard. Wilson's statement that his speakers used a marked trill suggests that his authorities may have been somewhat conservative in their speech, since Grant observes that "within recent years there has been a tendency to attentuate the force of the trill especially in final positions and before another consonant. . . . The trill may be reduced (finally and before consonants) to a single tap [ɾ], or even to a fricative consonant [ɹ]" (1913, 35).

There is another respect in which Wilson's sample seems to be conservative. Wilson makes no mention of glottal stops, although in a later work (Wilson 1926) he reports that the "glottal catch" is common in Fife. This suggests that Wilson would have mentioned the use of glottal stops if they had been prominent in the speech of his authorities. They are certainly common in the lower-class interviews and, according to Grant, were common "in Scotch dialect pronunciation in the district between the Firths of Forth and Tay on the east side and the Firth of Clyde on the west" (1913, 30).

The contemporary use of a glottal stop for /t/ is well established in the speech of various regions of the British Isles: London (Wells 1982), Norwich (Trudgill 1974), and Ulster (Gregg 1958). In Scotland there have been three

Table 4.1. Glottal Stops in Four
Environments

Category	#C	#V	##	V—V
Glasgow (Macaulay)	%	%	%	%
Adults	81	68	56	31
15-yr-olds	86	71	69	50
10-yr-olds	88	80	69	34
Edinburgh (Romaine)				
10-yr-olds	91	98	95	76
8-yr-olds	89	95	93	80
6-yr-olds	84	94	94	75
Edinburgh (Reid, 11-yr-olds)				
Interview style	97	55	62	48
Group style	100	69	69	61

empirical studies: one for Glasgow (Macaulay and Trevelyan 1973; see also Macaulay 1977) and two for Edinburgh (Romaine 1975 and Reid 1978). The Glasgow study included adults, 15-year-olds, and 10-year-olds; in Edinburgh Romaine investigated the speech of 6-year-olds, 8-year-olds, and 10-year-olds, while Reid reported on 11-year-old boys in two contextual styles. Romaine emphasized the importance of looking at different environments for glottal stops. She considered seven different environments, but for comparison only four will be considered here. The figures from the three studies are given in Table 4–1.

All three studies agree that the word-medial (V—V) environment is the least favorable for glottal stops. Reid and I found that the environment of word-final before a following consonant (#C) was the most favorable for glottal stops; but, somewhat surprisingly, Romaine found it to be less favorable than word-final before a pause (##) or a following vowel (#V).

This gives the ordering for the three studies as shown in Table 4–2.

Table 4.2. Environments Favoring the
Occurrence of Glottal Stops

Favorableness	Macaulay	Romaine	Reid
Most favorable	#C	#V	#C
↕	#V	##	##
	##	#C	#V
Least favorable	V—V	V—V	V—V

To investigate the use of glottal stops in the Ayr interviews, all examples of /t/ following a vowel in a posttonic syllable were identified in two contexts: before a vowel and before a pause. The primary reason for restricting the analysis to these two contexts was that these were the environments in which variability had been clearest in the Glasgow study, but it was also because these are the two environments in which the identification of the stop is easiest. As Romaine (1975, 115) points out, discrimination of stops before a following alveolar consonant is particularly difficult; and this tends to reduce the reliability of figures for the environment of a following consonant. No attempt was made to distinguish between the context of a pause and utterance-final position, because the category of utterance-final is difficult to establish even by means of intonation. The three environments, therefore, in which the use of glottal stops was tabulated are word-final before an immediately following vowel (#V), word-final before a pause long enough for the effect of the next segment to be negligible (##), and word-internal between vowels (V__V). The frequency of glottal stops in these three environments is shown in Table 4–3.

It can be seen from Table 4–3 that for both groups and for most individual speakers glottal stops are most frequent before a pause and least frequent

Table 4.3. Frequency of Glottal Stops

Respondent	#V	##	V__V	Total
Lower-class	% n	% n	% n	% n
Rae	83.3 (20/24)	95.2 (20/21)	76.9 (20/66)	84.5 (60/71)
Gemmill	78.9 (97/123)	71.2 (52/73)	67.8 (40/59)	74.1 (189/255)
Lang	58.6 (89/152)	85.7 (156/182)	49.2 (65/132)	66.5 (310/466)
Ritchie	46.2 (18/39)	56.1 (23/41)	40.0 (12/30)	48.2 (53/110)
Laidlaw	35.2 (32/91)	36.8 (32/87)	34.5 (41/119)	35.4 (105/297)
Sinclair	14.3 (28/196)	44.6 (74/166)	28.8 (38/132)	28.3 (140/494)
All LC	45.4 (284/625)	62.6 (357/570)	43.4 (216/498)	50.6 (857/1,693)
Middle-class				
Gibson	53.6 (60/112)	21.7 (5/23)	12.5 (8/64)	35.7 (73/199)
Nicoll	16.5 (26/158)	28.6 (18/63)	6.5 (8/124)	15.1 (52/345)
M'Dougall	12.1 (20/165)	26.5 (13/49)	1.2 (1/83)	11.4 (34/297)
Menzies	.0 (0/102)	8.0 (4/50)	.0 (0/60)	1.9 (4/212)
MacGregor	.0 (0/116)	7.4 (4/54)	.0 (0/56)	1.8 (4/226)
Muir	2.0 (1/51)	.0 (0/27)	.0 (0/17)	1.1 (1/95)
All MC	15.2 (107/704)	16.5 (44/266)	4.2 (17/404)	12.2 (168/1,374)

Table 4.4. Glottal Stops Morpheme-internal and
before an Inflection

Respondent	Morpheme-internal	Before an Inflection
Rae	75.0 (9/12)	78.6 (11/14)
Gemmill	60.0 (24/40)	84.2 (16/19)
Lang	25.6 (22/86)	93.5 (43/46)
Ritchie	5.9 (1/17)	84.6 (11/13)
Laidlaw	8.5 (6/71)	72.9 (35/48)
Sinclair	7.8 (7/90)	73.8 (31/42)
All LC	21.8 (69/316)	80.8 (147/182)

word-internally. The latter finding is consistent with the earlier studies; but the former is contrary to the order found in the Glasgow study and by Romaine in Edinburgh, though it supports Reid's results. Glottal stops are more than four times as frequent in the lower-class interviews as in the middle-class interviews. Among the lower-class speakers, however, there is also considerable difference in the frequency of glottal stops in these environments, with Rae using glottal stops three times as frequently as Sinclair. Ritchie, Laidlaw, and Sinclair form a group using less than 50% glottal stops in contrast to Rae, Gemmill, and Lang, who use more than 65% glottal stops. The word-internal environment was subdivided into two categories: (1) morpheme-internal and (2) before an inflectional morpheme. The results are shown in Table 4–4. (Only the figures for the lower-class speakers are given, since there were too few examples in the middle-class interviews to make quantification meaningful.)

The notable feature of Table 4–4 is the high proportion of glottal stops before an inflectional morpheme. The morphemes are past tense *–ed*, comparative *–er*, superlative *–est,* and inflectional *–ing*. For the purposes of this analysis *better* was treated as a suppletive form and not as containing the inflectional morpheme *–er*. Four examples of *meeting* used as a count noun were also excluded from the tabulation of inflectional *–ing*. It can be seen in Table 4–4 that for Ritchie, Laidlaw, and Sinclair the overwhelming majority of word-internal glottal stops occur before an inflection.

The majority of the inflectional morphemes, however, were *–ing,* 81.9% (149/182). Of these 149 instances, 142 (95%) were preceded by a glottal stop. For four speakers—Rae, Gemmill, Lang, and Ritchie—glottal stops were categorical before inflectional *–ing*. Since *–ing* is regularly represented by syllabic [n̩], it is possible that the occurrence of glottal stops is triggered by this feature. This would give a sequence as shown in (1).

(1)　a.　$\#/\text{ɪŋ}/ \rightarrow [\text{ŋ}̩]$
　　　b.　$/\text{t}/ \rightarrow [ʔ]/\_\_\_\_[\text{ŋ}̩]$

Support for this view comes from an examination of other words with syllabic [ŋ̩]: *Brighton, Britain, Latin, retina, rotten*, and *threaten* all occur with syllabic [ŋ̩] and all are preceded by glottal stops; whereas *British*, for example, occurs only with [t]. (The 17 examples of word-medial glottal stops in the middle-class interviews are also before syllabic nasals.)

An examination of alveolar nasals in the following word, however, suggests that it is the alveolar nasal rather than its syllabic quality that favors glottal stops. There are 49 instances where a syllable ending in /t/ is followed by a word beginning with a vowel and /n/, for example, *get in, right enough, sat on*, and so on. In 44 cases (89.8%) the preceding /t/ is realized as a glottal stop. Moreover, three of the five exceptions are clearly marked by the kind of syllable division mentioned by Abercrombie as characteristic of Scottish English: "Scottish speakers make as many syllables open as possible (an open syllable is one that ends in a vowel), so that they will attach the final consonant of a word to the beginning of the next word, if it starts with a vowel" (1979, 82). Abercrombie's example is [sən ˈtændruz] for *St. Andrews*. This kind of syllable division is very common in the interviews and has the effect of converting a syllable-final consonant into an initial consonant. For example, there are 23 instances of *at all* in the interviews; and there is no possibility of a glottal stop in this environment because the /t/ is in pretonic position. In three out of the five cases where final [t] occurs before a following vowel plus /n/ the syllable division is very clear (e.g., [gɔ tʰɔn] for *got on*), and this accounts for the absence of a glottal stop. It can, therefore, be claimed that there is a very strong tendency—categorical for some speakers—favoring glottal stops before /n/, whether syllabic or preceded by a vowel.

On the other hand, following liquids tend to disfavor glottal stops. /r/ disfavors the occurrence of glottal stops as does /l/, though not as strongly (____Vr 13.8% [19/138], ____Vl 33.3% [11/33]). It is perhaps not a coincidence that the most commonly cited examples of stigmatized glottal stops include *butter, water*, and *better* (e.g., McAllister 1963, 66), since a glottal stop will be highly salient in this environment.

There are, however, two environments that even more strongly disfavor glottal stops. One is when /t/ is followed by V + /t/ in the next syllable. In a kind of consonant harmony both examples of /t/ are usually realized as [t]. For three of the speakers (Ritchie, Laidlaw, and Sinclair) this rule seems to be categorical; in the interviews with Rae and Gemmill there are too few clear examples to be sure. In Lang's interview, however, there are 24 examples of /t/ before V + /t/. In 18 instances (75%) the sequence is [t]V[t]; in one case the first /t/ is flapped [pɛ̈ɾɛ̈t] *put it;* in three instances the sequence is [ʔ]V[ʔ], twice before a consonant in the next syllable and once before a pause. Since glottal stops are more frequent before a consonant or a pause, the first glottal stop may be a kind of anticipatory assimilation. In two instances, the second /t/ is followed by a vowel plus /n/. In these cases, the first /t/ is realized as [t] and the second as [ʔ]: [p ɛ̈t ɛ̈ʔ ɛ̈n] *put it in:* [nəɪt ʌʔuən] *night out and.* This suggests that the rule favoring glottal stops before /n/ operates after the rule that favors [t] before V + /t/.

The second environment that favors [t] is before the hesitation marker *eh* [ɛ]. There are 52 instances of /t/ before /t/ before *eh* (45 of them *but*), and in every case /t/ is realized as [t]. Since glottal stops are frequent before a pause, the use of [t] in this environment is presumably a signal that the speaker has not finished and wishes to continue the turn.

Finally, there is one category of voiceless alveolar stops that is never realized as a glottal stop. Inflectional *–ed* is frequently devoiced even when not preceded by a voiceless sound, for example, [wantɛ̈t] *wanted.* Such stops are never realized as glottal stops, so either the rule for devoicing must apply after the rule for glottalization or the latter rule must be blocked in some other way.

The use of glottal stops for oral stops is often stigmatized and characterized as sloppy or careless, particularly by teachers (McAllister 1963, 71; Macaulay and Trevelyan 1973, 158). The preceding analysis shows that the occurrence of glottal stops in the Ayr interviews is by no means random but follows a predictable pattern. Like any other linguistic variable, the use of glottal stops in Ayr displays structured heterogeneity.

The velar fricative /x/ is found in both the middle-class and the lower-class interviews. Its distribution, however, is somewhat different in the two groups. In the middle-class interviews it occurs only in proper names (e.g., *McCulloch, Auchinleck,* etc.) and in the discourse marker *och.* Three of the lower-class speakers use /x/ more widely. The most significant of these occurrences are in alternation with ø for orthographic *gh* as in [nɛ̈xt] versus [nəɪt] for *night.* Rae uses /x/ 14% (2/14) of the time, Lang 38% (16/42), and Gemmill 43% (30/70). Rae uses /x/ only in *night* (though he also uses [nəɪt]). Lang uses a velar fricative in *brought, eighty, Kirkmichael, might, night, ocht, richt, tightened* [tɛ̈xənd], and *weight.* Gemmill has /x/ in *bought, daughter, eight(th), eighteen, fright(en), light, might, night, right, thought, tight,* and *wrocht.* This is a shorter list than Wilson gives, but it is extensive enough to show that at the time of the interviews the velar fricative was still widely used by some speakers in the community and not restricted to a few residual forms such as *dreich* [drɪx].

Wilson records various examples of consonant deletion that are also found in the lower-class interviews. The first is the deletion of a homorganic voiced stop after a nasal. The following examples occur in the interviews with Gemmill, Lang, Rae, and Sinclair:

(2) a. [bɔn] *band*, [dɔnər] *dander*, [grʌn] *ground*, [hɔnz] *hands*, [hɔnəl] *handle*, [hɛnərsən] *Henderson*, [hʌnər] *hundred*, [run] *round*, [sanë] *Sandy*, [stɔnën] *standing*, [wʌnər] *wonder*, and [yɔnər] *yonder*

b. [gamlën] *gambling*, [rʌməl] *rumble*, and [tʌmlər] *tumbler*

c. [aŋrë] *angry*, [ëŋlʌn] *England*, [mʌŋrəl] *mongrel*, [sıŋəl] *single*, [swıŋəltri] *swingletree*

Homorganic voiced stop deletion is not a categorical rule for any of the male speakers, though it occurs regularly in all four interviews. Laidlaw has two examples, Ritchie has none.

The next case is the deletion of /l/ under certain conditions. These are survivals of the fifteenth-century vocalization of /l/ (Aitken 1985, xiv).

(3) a. [ɔ] *all*, [bɔ] *ball*, [kɔ] *call*, [fɔ] *fall* (v. and n.), [hɔ] *hall*, [wɔ] *wall*

b. [hɔd] *hold* (v. and n.)

c. [rʌu] *roll*

d. [pu] *pull*

e. [sodʒər] *soldier*

f. [fɔt] *fault*

With the exception of the forms in (3a) which occur variably throughout the lower-class interviews in alternation with forms in which /l/ is retained, this phenomenon is restricted to a small number of lexical items. There are also words containing /ɔl/ that do not show vocalization, for example, *always*, *auld*, *bald*, *cauld*, *crawl*, *doll*, and *haul*. There are other signs of lexical differentiation. *Holding* as a participle occurs with vocalization of /l/ but *holding* in the sense of "farm" does not. On the other hand, there may be contextual constraints on vocalization, since no examples are found before inflectional −*s*, though the numbers are very small. A different kind of constraint is shown in the word *football*: there are 19 examples of this item, 11 with vocalization of /l/. In all of the cases with vocalization the first vowel is [ë]; in the remaining 8 cases without vocalization the first vowel is [u]. For the most part, however, the variation seems to be unconstrained in the items in which vocalization most commonly occurs, that is, *all, call*, and so on. On many occasions throughout the lower-class interviews a form with vocalization will be closely followed or preceded by the same item without vocalization, as shown in (4):

(4)   a. AS1916   well maybe that was *aw*
          1917   you spent on beer *all* night
      b. WL2159   of coorse it's a big *hall* that
          2160   there no doubt aboot that
          2161   it's a big *haw*
      c. WL1845   and they went *all* over the place to the hervest big
                  mills tattie planting tattie weeding onything at *aw*
      d. WL2364   so that's the eight of them
                  (*All* accounted for)
          2365   they're *aw* accounted for aye

Even the interviewer's use of *all* does not prevent vocalization in (4d). Nor is there any indication of effect of topic or genre on vocalization, as it is distributed fairly evenly throughout the interviews. The proportion for individual speakers is given in Table 4–5. This table shows a distribution similar to that of earlier tables, with Sinclair, Ritchie, and Laidlaw forming a group with less than 50% vocalization in contrast to over 65% for the others.

Wilson reports a number of cases in which medial and final /v/ is deleted, but only five of these occur in the lower-class interviews: [o] *of*, [gi] *give*, [he] *have*, [li] *leave*, and [ʌuər] *over*. Other consonantal forms that are found only in the lower-class interviews include [ɔðəgëðər] *altogether*, [brëg] *bridge*, [pëk] *pitch*, [skaʳtët] *scarted* (i.e., scratched), [sklëmën] *sclimming* (i.e., climbing), [smëdë] *smithy*, and [ʃënər] *sooner*.

Wilson states that −*it*/−*t* is the normal form of the past tense and past participle morpheme, though he admits that −*d* occurs variably for some speakers with certain verbs that end in a vowel and is the obligatory form with other verbs. The occurrence of [ët/t] as the past tense or past participle morpheme after voiced sounds is much more restricted in the Ayr interviews. [t] occurs after a vowel only with disyllabic verbs where the second syllable is

Table 4.5. Vocalization of /1/

Respondent	Percentage
Rae	91 (31/34)
Gemmill	85 (55/65)
Lang	66 (58/88)
Laidlaw	47 (33/70)
Ritchie	40 (17/42)
Sinclair	37 (55/147)
All LC	56 (249/446)

/r/ + V, for example, [bërët] *buried*, [kerët] *carried*, [merët] *married*, and [wʌrët] *worried*. [t] also occurs frequently after nasals and liquids, for example, [blemt] *blamed*, [bəlɔŋt] *belonged*, [klint] *cleaned*, [këlt] *killed*, and [purt] *poured*. [t] also frequently occurs after a final /v/ with anticipatory devoicing to [f], for example, [bəlift] *believed*, [ëmpruft] *improved*, [lëft] *lived*, and [sɔlft] *solved*. [ët] occurs only after verbs ending in /t/, never after verbs ending in /d/, for example, [startët] *started* and [wantët] *wanted* but [landëd] *landed* and [paredëd] *paraded*.

Vowels

Wilson does not give descriptions of the vowels, but he provides IPA equivalents for his transcriptions. Although this does not provide all the information that one would like, it is sufficient to establish the basic system. In Wilson's account and in the later interviews the basic system is the same: nine vowels and four diphthongs:

> *Vowels:* i, ɪ e, ɛ, a, ʌ, ɔ, o, u
> *Diphthongs:* ae, əɪ, ʌu, ɔe

/i/ is a high front unrounded vowel, fairly peripheral, close to CV 1, with little off-glide[1]. It is frequently very short. This vowel is used variably by lower-class speakers in alternation with other vowels in a small number of words: with /ɪ/ in *drill, finish* (first syllable), *gie, minute* (first syllable), *sick,* and *women* (first syllable); with /ɛ/ in *beheaded* (second syllable), *dead, eleven* (second syllable), *head, Heads of Ayr* (first word), *instead* (second syllable), *medals* (first syllable), *seven* (first syllable), *stretcher* (first syllable), *sweat,* and *well* (adverb); with /e/ in *neighbor* (first syllable), and *swear;* and with /ae/ in *die*

/ɪ/ varies in phonetic quality from a slightly centralized high front unrounded vowel to a fairly low, slightly retracted unrounded central vowel. It is mainly the middle-class speakers who have a fairly front vowel, particularly before /ʃ/ and /ŋ/; but even the middle-class speakers frequently have a centralized and/or lowered vowel before /r/, /v/, and /m/. There is no neutralization with /ɛ/ and /ʌ/ before /r/; that is, *bird* [bɪrd], *heard* [hɛrd], and *word* [wʌrd] do not rhyme. In the lower-class interviews /ɪ/ is often quite low and retracted before /l/ and /r/. Three of the lower-class speakers—Gemmill, Lang, and Rae—variably use /ɪ/ for the short reflex of Old Scots /o/ in *afternoon, foot, football, good, school, soon, wood,* and *wooden.* It also occurs in *used* [jëst] (in its quasi-modal function) and in *put.* Sporadically, /ɪ/ occurs for /ɛ/, as in *get* and *yet.*

[1]CV refers to the Cardinal Vowel system developed by Daniel Jones.

/e/ is a half-close front unrounded vowel close to CV 2, with little or no off-glide. One of the most salient features of lower-class speech is the use of /e/ for /o/ in word such as *both, clothes, floors, go, home, more, most, mostly, no, nobody, none, own, sore, stone, whole.* In the lower-class interviews /e/ is also frequently, but not categorically, used in words containing the sequence /ar/ + c and in a few other words: *apple, Archie, arm, card, carry, charge, Charlie, family, father, garden, Gardener, guard, harm, jacket, Kilpatrick, married, master, matter, part, rather,* and *yard.* /e/ also occurs in *beat, seat,* and in *do, too, sure,* and *surely.*

/ɛ/ is a half-open front unrounded vowel close to CV 3; it is frequently lengthened in emphatic syllables. In the lower-class interviews /ɛ/ variably alternates with /a/ in certain words: *after, afternoon, class, farm, farmer, glad, harvest, marched, Saturday, start,* and *travel.* /ɛ/ also occurs in *game.*

/a/ is a low unrounded vowel that ranges from slightly higher than CV 4 all the way back to CV 5 with the most common variants occurring in the low central region. In the lower-class interviews /a/ alternates with /ɔ/ in a few words, mostly before labials or following /w/: *drop, off, top, wrong;* and *swallowing, wallop, want, wash, washhouse,* and *water.*

/ɔ/ is a low back rounded vowel closer to CV 5 than to CV 6. This is the vowel used by some of the lower-class speakers in alternation with /a/ as the back variant in the words *band, car, dander, dark, far, half, hand, handle, stand,* and *tar.* It also occurs occasionally in alternation with /o/ in *hold* (verb and noun with loss of /l/), *row,* and *snow.* It also occurs variably as the second vowel in *away.*

/o/ is a half-close back rounded vowel close to CV 7. In the lower-class interviews there is frequent neutralization between /ɔ/ and /o/, particularly before stop consonants, and among these especially voiceless stops, for example, [blok] *bloke* and *block,* [not] *note* and *knot.*

/u/ is a high central rounded vowel about midway between CV 9 and CV 8. /u/ alternates invariably with /ʌu/ in the lower-class interviews. /u/ occurs in the following words: *about, account, allow, around, bowels, Brown, council, councillors, count, county, crown, down, encounter, found, here-abouts, house, how, Howie,* (Minnie) *Mouse, mouth, now, nowadays, our, out, out-of-date, outside, plough, Portsmouth, round, south, stour, thousand, town, trousers,* and *without.*

/ʌ/ is a half-open centralized back unrounded vowel. It is used variably by some of the lower-class speakers in a few words: *bull, look* (the verb but not the noun); *body, clogs, dog, donkey, not; ground, pound* (weight); *three-pence.*

/ae/ is a relatively long diphthong starting from a low central position and rising to a position slightly above half-close front. It occurs in the "long" environments of the Scottish Vowel Length Rule (Aitken 1981).

/əɪ/ is a centralized diphthong beginning about a midcentral position and rising to a fairly high front position. It is used in the "short" environments of the Scottish Vowel Length Rule. This diphthong is used variably in the lower-class interviews in words with the reflex of Old Scots /ai/: *gey, hay, jail, May* (month), *Maybole, paid, pay, stay, way.* /əɪ/ also occurs in *boyne* ("washtub") *join, joiner,* and *quoiting;* and in *change.*

/ʌu/ is a relatively short diphthong starting from slightly below a midcentral vowel and rising to a high central rounded position. It is frequently so short as to be difficult to distinguish from monophthongal /u/, with which it alternates in the lower-class interviews. In the lower-class interviews an open variety of /ʌu/ alternates with /o/ in *four, grow, howed, howk, roll, soul,* and *Tarbolton.* /ʌu/ also occurs in *coup, loup,* and *lowse.*

/ɔe/ is a relatively long diphthong beginning as a half-open rounded back vowel about midway between CV 6 and CV 7 and fronting at about the same height to a half-open unrounded vowel slightly above CV 2. It occurs in words such as *boy, enjoy,* and *noise;* it alternates with /əɪ/ in *join* and *joiner.*

The Variable (au)

In Middle English and Old Scots the corresponding vowel was a high back rounded vowel /u/. In most English varieties this has become a wide rising diphthong /au/; but in northern dialects, including lowland Scottish dialects, a high rounded monophthong has survived. In both the diphthong and the monophthong there is considerable variation in the actual phonetic quality, so that this variation can be treated as a continuum (Macaulay 1977). For the purposes of the present discussion, however, phonetic quality is not the issue; and the variants can be classed as either diphthongal or monophthongal.

In the Ayr sample, there is a clear division between the middle-class speakers, who always use a diphthongal form, and the lower-class speakers, who frequently use a monophthong. However, none of the lower-class speakers always uses a monophthong, as can be seen from Table 4–6. Since the interviews differ greatly in length and somewhat in topic, the proportion of each form used is what counts. It can be seen from Table 4–6 that all the lower-class speakers use some diphthongal form; but the frequency varies greatly, and there are two clear subgroups. Sinclair, Ritchie, and Laidlaw again form a group, using less than 50% monophthongs; Rae, Gemmill, and Lang form another group, with a frequency of more than 80% monophthongs.

Table 4.6. The Variable (au): Total
Interview

Respondent	Monophthong		Diphthong	
	%	n	%	n
Lang	87	217	13	31
Gemmill	85	191	15	34
Rae	83	68	17	14
Laidlaw	48	96	52	103
Ritchie	39	31	61	49
Sinclair	31	112	69	249
All LC	60	715	40	480

It is possible that the two groups do differ to this extent in their use of the variable (au) in their normal everyday life despite the fact that (with the exception of Gemmill) they all live in the same district and know each other. But it is also possible that the difference between the two groups reflects a difference in sensitivity to the presence of an interviewer who used only diphthongal forms. Some support for this view can be found in Table 4–7, which shows the proportion of monophthongal forms in the first and second halves of the interviews. It can be seen from Table 4–7 that for four of the speakers, there is a clear increase in the use of monophthongal forms as the interview progresses. This suggests that they relax somewhat as the interview proceeds and that their use of diphthongal forms is at least partly a response to the interview situation. However, Table 4–8 shows evidence for another factor, which can be clearly seen in the interviews with Sinclair and Laidlaw. In the first half of each interview is a narrative section dealing with an early

Table 4.7. The Variable (au): Change
between Halves of the Interview

Respondent	Monophthongal Forms (%)		
	1st Half	2d Half	Change
Gemmill	89	81	−8
Lang	88	86	−2
Rae	73	97	+26
Laidlaw	38	51	+13
Ritchie	31	52	+21
Sinclair	26	36	+10

Table 4.8. The Variable (au) in
Narratives

	Monophthongal Forms (%)	
Respondent	Narrative	Elsewhere
Sinclair	82	29
Laidlaw	71	46

experience: in telling of this experience, each speakers makes a much greater use of monophthongal forms. Since these narratives occur relatively early in the interviews, it is clear that what is affecting the frequency of monophthongal forms here is genre, not adaptation to the interview situation. It is also significant that the shift should be so great in the case of Sinclair, since (as will be shown later) his speech differs in many ways from that of the other lower-class speakers.

A factor that can affect the frequency in the other direction is that certain words are either categorically diphthongal or have a greater likelihood of being diphthongal. For example, *shout* is invariably diphthongal, perhaps to avoid homonymity with *shoot*. *Pound* (sterling) is always diphthongal, though *pound* (weight) can be monophthongal [pʌn]. *How* is frequently—but not categorically—diphthongal, perhaps because of a possible confusion with interrogative *who*. Where there is no possibility of confusion, *how* can be monophthongal (e.g., Lang [hu ʃoːd jə hu tə spləis ðə rops], "Who showed you how to splice the ropes?"). There is also a tendency for *now* as a discourse marker (see Chapter 10) to be diphthongal in contrast to *now* as a temporal adverb, which is frequently monophthongal. Less common words are more likely to be diphthongal. For example, in the interviews with Gemmill and Lang, the following words occur only with diphthongs: *accounted, amount, howling, northbound, southbound, Southsea,* and *underground* (although *count* and *ground* occur with monophthongs, [kunt] and [grʌn]). It is the relatively high proportion of these rarer words in the second half of the interviews with these two speakers that accounts for the slight increase in diphthongal forms there. It is clear that monophthongal forms are normal for Gemmill and Lang in common words; but even then, a diphthong is available for stylistic emphasis or rhetorical purposes. For example, in Lang's interview *out* is almost categorically monophthongal; but in describing an accident down in the mine, he uses diphthongal forms twice in quick succession:

(5)   WL897   and yin came up
          898   and hit him just on the side of the heid
          899   [psssst] *out* like a light
          900   *out* like a light

Similarly, for Lang, *down* is regularly monophthongal; but in talking about the minister who was opposed to gambling, he twice uses the diphthongal form:

(6)   WL2895   on the one hand he was *down*
                 2896   on gambling
                 2897   but he wasnae *down* on drink
                 2898   because I ken that
                 2899   I poured it oot for him

Here it is as if the metaphorical use of *down* has made it into a different item. This example occurs very nearly at the end of the interview and cannot be a matter of accommodating to the interviewer. A similar example occurs late in the interview with Gemmill. Throughout the interview *out* is categorically monophthongal except for one instance when he is describing the un-availability of seats on the train:

(7)   HG1451   so I went back to Euston
                 1452   to book saits back to Glesca
                 1453   oh they're *out*

The use of the variable (au) illustrates the effect of a number of factors on the choice of form and shows that the variation does not depend on a single stylistic dimension.

The Variable (e)

Unlike the variable (au), in which there is a continuum between diphthongal and monophthongal forms, the variable (e) is discontinuous, with *home* (for example) being either [hom] or [hem] and there is never any doubt as to which form is being used. Since it is the use of /e/ that is salient, items from different sources have been grouped together. Only items in which there is alternation with /e/ have been included. For example, *too* in the sense of "also" occurs as [te], but *too* as an intensifier with the meaning "excessively" never does. Consequently, only instances of the first use are tabulated. Two other arbitrary exclusions have been made. All forms of the verb *go* have been excluded even though two examples of [ge] were recorded (one each in the interviews with Lang and Rae). Since there were approximately 250 examples of *go,* to have included them would have distorted the picture. In Sinclair's interview there were 15 examples of the hedge *more or less;* and they were also omitted, since there is no evidence that they participate in the alternation [mor]/[mer]; if they had been included, they would simply have reduced

Table 4.9. The variable (e) (%)

Respondent	Percentage
Rae	90.9 (60/66)
Gemmill	88.8 (79/89)
Lang	81.2 (82/101)
Laidlaw	40.7 (35/86)
Sinclair	19.5 (22/113)
Ritchie	8.1 (3/37)
All LC	57.1 (281/492)

Sinclair's already low percentage by another two percentage points. Table 4–9 shows the percentage of /e/ for the six lower-class speakers. Not surprisingly, the pattern is similar to that of (au), with Rae, Gemmill, and Lang forming a group with over 80% /e/ and Laidlaw, Sinclair, and Ritchie another group with less than 50% /e/. Again, there are indications in the interviews with Laidlaw and Sinclair that genre and topic affect the frequency of /e/, with narratives and jocular asides being more likely to contain it. Laidlaw does not use any /e/ forms before her husband enters the room; he comes in at line 448 and the first /e/ form occurs at line 473.

The Variable (a)

The class consists of the items that—categorically in the middle-class interviews and variably in the lower-class interviews—have /a/. The middle-class speakers have a low central vowel [a] that may be advanced or retracted somewhat depending upon the surrounding sounds, but it does not become either a front or a back vowel. The range of /a/ for the lower-class speakers (most noticeably in the interviews with Rae, Gemmill, and Lang) is shown in Table 4–10. This is the kind of distribution found by Milroy (1982) in Belfast, as shown in Table 4–11. Milroy, although cautioning, "We do not necessarily know beforehand what is the correct lexical input to any variable" (1982, 39), treats this range of pronunciations as phonologically conditioned variation. Trudgill suggests that it is rather a case of dialect mixing, in which "different phonological variants present in a dialect mixture situation, and not levelled out during focusing, may be retained as *allophonic variants*" (1986, 125, emphasis original). In the Ayr situation there is no reason to suggest that the distribution results from dialect mixing, and it is not clear that all the variants can be explained by phonological conditioning. [ɔ] before /r/# and /nd/ (with

Table 4.10. The Variable (a)

[ë]	[e]	[ɛ]	[a]	[ä/ä́]	[ɔ]
hang*	apple*	after*	swallow*	are*	band*
have*	Archie*	afternoon*	wallop*	army†	car*
	arm*	class*	want*	carpets	dander*
	card*	farm*	wash*	gamble	dark*
	carry*	farmer*	water*	hard*	far*
	cart*	glad*		lassies*	half*
	charge*	harvest†		mark*	hand*
	Charlie*	marched†		narrow†	handle*
	father*	Saturday*		pass*	stand*
	family*	start*		shaft*	tar
	garden*	travel*		traffic	
	Gardener				
	guard*				
	harm†				
	jacket*				
	Kilpatrick				
	married*				
	master*				
	matter*				
	part*				
	party*				
	rather				
	yard*				

*consistent with Wilson.
†inconsistent with Wilson.

the /d/ assimilated to /n/ or elided) might be phonologically conditioned, but that would leave *dark* and *half* unexplained. [a] after /w/ is the historically unrounded vowel in northern dialects, though it is interesting that the vowel should be more advanced than the basic vowel [ä/ä́]. The most complex situation, however, is the environment before /r/ followed by a consonant or a vowel. We find [e] in *Archie**, *arm**, *card**, *carry**, *cart**, *charge**, *Charlie**, *garden**, *guard**, *harm+*, *married**, *part**, *party**, and *yard**, while [ɛ] occurs in *farm**, *farmer**, *harvest+*, *marched+*, and *start**. (The items marked with an asterisk are those that are reported by Wilson as having the same vowel, and the plus indicates that Wilson records the other vowel.) Omitting *harvest* (since Wilson gives only *hairst*), which makes the comparison more disputable, there are 16 out of 18 cases (89%) where the two

Table 4.11. Range of /a/ for a Working-Class Belfast Speaker

[ɛ]	[æ]	[a]	[a]	[ɔ]
bang (5)	Bangor	Castlereagh (2)	bad	past
crack (2)	jacket	barracks	Strathearn	plaster
Kojak	back	that (3)	have	palace
avenue	cracking	snap	wrap	hand
		cracking	happen	Belfast (2)
		Albert	Catholic (2)	handsome
		avenue	Castlereagh	can
		baton (2)	Belfast	strand
		camera	happy	
		Shankhill (2)	handsome	
			candid	

Milroy 1982, 41

accounts agree in the distribution of the vowels. (For the two words that differ, *harm* and *march,* the *Concise Scots Dictionary* [Robinson 1985] gives both [e] and [ɛ].) Since the phonetic difference is small and does not mark a phonological contrast, this is a situation in which one might have expected merger, or at least confusion. There is no sign of either, though this may be less surprising in a variety that generally keeps vowels distinct before /r/. The distribution suggests that the variation is lexically determined rather than phonologically determined. This constitutes evidence for a stable situation rather than one in which the variation is the result of linguistic change in progress.

I have shown that there are clear differences between the middle-class speakers and the lower-class speakers in their use of both consonants and vowels. The lower-class phonological system is much more complex than that of the middle class. This is consistent with what James Milroy found in Belfast:

> Working-class speech (even when it is careful speech) is rich in low-level phonological variation: middle-class speech appears to suppress this variation. While it would be premature to claim that this is a universal, I believe it will turn out to be generally true of large urban speech communities where there have been long histories of dialect diversification in rural hinterlands and recent urban growth. It may be less easy to demonstrate this pattern in long-established urban dialects or in rural forms. (1982, 46)

With a population of less than 50,000 Ayr is marginally a "large urban speech community," but there is no evidence of "dialect diversification" in the rural hinterlands. On the contrary, the evidence from Wilson suggests that the phonological variation has been stable. The existence of variant forms allows the lower-class speakers a stylistic choice that is not available to the middle-class speakers (except in mimicking lower-class speech; see Chapter 11).

It would be misleading to attribute this variation to the influence of a single factor such as attention to the speaker's speech (Labov 1966a), formality (Coupland 1988), audience design (Bell 1984), or speech accommodation theory (Thakerar, Giles, and Cheshire 1982). As Levinson has pointed out, the problem is how to define a notion of *style* "that will capture patterns of language use that are not part of a dialect but are typical and distinctive of speakers of a dialect" (1988, 164–65). For example, it would be absurd to claim that Lang and Gemmill normally speak a "vernacular" in which the realization of *out* is [ut] but that on occasions (for reasons of formality, recipient design, or speech convergence) they "code-switch" to a "Standard English" realization [aut]. Rather, both pronunciations are part of their language/dialect/competence/speech repertoire; and the choice of one or other will be governed by a number of factors, not all of which are dependent upon interactional features.

Levinson also draws attention to a problem with the notion *sociolinguistic alternate:* "A sociolinguistic alternate is, of course, one of two [or] more linguistic forms that are in paradigmatic opposition and *express the same meaning* or at least *perform the same function*" (1988, 165, emphasis original). The problem arises because it is not always obvious when two variants are synonymous (Lavandera 1978). For example, it seems safe to claim that [dun] and [daun] are not always synonymous in Lang's speech (see Item 6), but what about *all* and *aw* in Sinclair's and Lang's speech (see examples in Item 4)? The alternation between the two forms has no obvious explanation in terms of formality, attention to speech, or accommodation and has no apparent semantic significance; but does that mean that it is meaningless? (There is a slight tendency for vocalization of /l/ to be inhibited by a following vowel, but the immediate phonetic context does not provide an adequate explanation of the greater part of the alternation.)

In the absence of more detailed information about the linguistic context in which variables occur, it may be wise to treat conclusions drawn from quantified indexes with some caution.[1] This does not necessarily mean that the patterns revealed by such indexes do not point to something real in the use of language by members of the community. On the other hand, some of the

---

[1]In the final analysis, the notion that variation is conditioned by external factors is rather demeaning. It is almost as if the investigator were saying, "Poor things, they know not what they do; but because I am a linguist, I know what forces are shaping their speech." I believe that speakers often know what they are doing in *choosing* among variants and that they are far from passive vehicles for external forces.

claims about linguistic changes in progress may turn out to have been made prematurely on the basis of simplistic notions of the motivation for a choice between alternates. Time will tell how accurate the predictions have been, but in the meantime careful attention to the context in which variants occur will help to avoid some of the dangers of quantification.

5

Three Salient Variables: Negation, Verb Concord, and Relative Clauses

Negation

Apart from the kind of phonological differences described in the previous chapter, negation is the feature that shows the most salient social class differentiation. There is a clitic form of negative found with most but not all auxiliaries that will be represented orthographically as *–nae*, although the pronunciation varies [ne], [nɛ], në], and [nʌ]. As can be seen from Table 5–1, this is the preferred form of clitic negative for all the lower-class speakers. (The only middle-class speaker who uses *–nae* is Nicoll, who has 9 instances out of 72 examples (12.5%), four of them in quoted direct speech and the others in narrative contexts.) All the lower-class speakers use *don't* categorically (as do the middle-class speakers, of course). Without *don't*, Gemmill and Rae categorically use *–nae*, and the other lower-class speakers use–*n't* very rarely. There are no examples of **isnae* or **arenae*. This is perhaps less surprising, since *isn't* and *aren't* are rare even in the middle-class interviews. Out of a possible 52 instances of *isn't* in the middle-class interviews, there are only 15 examples, 8 of them from Gibson and 5 of them from MacGregor. *Aren't* is even rarer, with only 2 examples out of a possible 20 instances.

Middle-class speakers tend to avoid forms such as *isn't, aren't,* and (to a lesser extent) *won't* by cliticizing on the subject rather than on the auxiliary, as in the examples in (1). Out of a total of 125 instances 108 (86%) are cliticized on the subject and only 17 (14%) on the auxiliary.

Table 5.1. Negative Clitics: Lower-Class Sample

Word	HG	EL	WL	WR	MR	AS	All
−n't							
don't	17	36	28	9	19	52	161
can't	0	4	0	0	3	2	9
isn't	0	3	3	0	1	0	7
haven't	0	2	0	0	0	0	2
daren't	0	1	0	0	0	0	1
Total	17	46	31	9	23	54	180
−nae							
didnae	32	31	30	6	10	15	124
doesnae	1	9	3	1	2	12	28
hadnae	4	5	5	0	6	4	24
hasnae	0	2	2	0	0	1	5
havenae	0	1	3	1	5	1	11
wasnae	12	20	23	3	10	10	78
werenae	16	3	2	5	2	4	32
cannae	3	4	7	6	3	4	27
couldnae	12	15	13	6	3	12	61
darenae	0	2	0	0	0	0	2
shouldnae	1	1	1	0	0	1	4
wouldnae	5	25	10	6	4	3	53
Total	86	118	99	34	45	67	449
%−n't	17	28	24	21	34	45	40
%−n't w/o *don't*	0	8	3	0	8	3	4

(1) a. NM2044   *that's not* put to the test
     b. DN616   "you north people *you're not* so hot"
     c. WG907   *there's not* the same sense of class at all
     d. DN370   and *he'll not* come in

In some cases where cliticization on the subject is not possible, no contracted form is used; although cliticization on the auxiliary would be possible, as in the examples in (2):

(2) a. JM2001   who *are not* prepared
        2002   to do a thing for themselves
     b. DN1062 and Largs *is not* so far away

    c. WG1466   some of them *are not* profound
    d. DN171   which *is not* permitted

Occasionally, uncontracted forms are found even when cliticization on the subject would be possible:

(3)   a. DN594   "I *will not* have it"
      b. DN181   "Dear Mr Nicoll you *will not* remember the incident"

The lower-class speakers also cliticize on the subject of the verb *be*. Out of 205 instances, 198 (97%) are cliticized on the subject, and only 7 (3%) on the auxiliary. The lower-class speakers frequently use *no* instead of *not* as the uncontracted form of the negative, as in the examples in (4):

(4)   a. WL457   "*I'll no* be so early the morn Bobby"
      b. AS737   *that's no* exaggerating
      c. EL3173   she *was no* as big as me
      d. WR395   ach *there no* much difference in the pits

The frequency with which the lower-class speakers use *no* in this function can be seen in Table 5–2. It can be seen from Table 5–2 that all the lower-class speakers, with the exception of Sinclair, use *no* more than half the time. The difference between Sinclair and the other speakers suggests that this feature is sensitive to monitoring. None of the middle-class speakers uses *no* in this way.

This use of *no* must not be confused with the use of *no* as a quantifier. In

Table 5.2. *No* Versus *Not*: Lower-Class Sample

	No		Not	
Respondent	*n*	%	*n*	%
Rae	26	93	2	7
Lang	52	85	9	15
Gemmill	23	74	8	26
Ritchie	12	71	5	29
Laidlaw	32	62	20	38
Sinclair	21	36	21	64
All LC	166	67	81	33

Table 5.3. *Nae* Versus *No*: Lower-
Class Speakers

Respondent	*Nae*		*No*	
	n	*%*	*n*	*%*
Gemmill	21	91	2	9
Rae	13	76	4	24
Lang	22	52	20	48
Sinclair	5	26	14	76
Laidlaw	2	14	12	86
Ritchie	0	0	9	100
All LC	63	51	61	49

this function the lower-class speakers, particularly the men, often use *nae*, as illustrated in the examples in (5):

(5) a. WL342   although I had *nae* matches
     b. WR081   well I had *nae* father
     c. AS3141   I mean there were *nae* doot aboot it
     d. HG1671   you hadnae *nae* mair money

The frequency with which *nae* is used in this function is shown in Table 5–3. The use of *no* in the lower-class interviews is consequently ambiguous. It may correspond to *not* in the middle-class interviews, but it may also be the same as the middle-class use of *no*. There is one example of *nae* in Gemmill's interview where *no* might have been expected: HG741 *she was nae that auld at the time*.

With the exception of question tags, cliticized forms of the negative do not occur in interrogative sentences, either in the lower-class or the middle-class interviews. There are only 13 instances of interrogative negatives, illustrated in (6):

(6) a. WL2269   "Was it no Ayala?"
     b. EL1241   "How are you no going back in?"
     c. HG171   did I not think
        172   I had been long enough firewatching
     d. WG937   "Could we not get expenses?"
     e. DN1881   why are they not getting on with their business?

There are only 17 examples of question tags, 9 in the lower-class interviews and 8 in the middle-class interviews. Fourteen of the question tags are negative, 10 of them *isn't*. It is interesting (though the numbers are too small to be significant) that 8 of the tags (47%) are used by women, although the interviews with the women make up only 18% of the whole. This is consistent with Lakoff's view that women are more likely to use question tags than men (1973, 55).

There is a greater tendency in the lower-class interviews to use negative incorporation. English allows two possibilities for the combination of a negative with an indefinite, as shown in (7):

(7)

	not incorporated	*incorporated*
	neg any	no/none
	neg anybody	nobody
	neg anywhere	nowhere
	neg anything	nothing
	neg ever	never
	neg either	neither

According to Quirk and his colleagues, the unincorporated form is more colloquial and idiomatic, with the possible exception of *never* (1972, 377). In the Ayr interviews, there are 522 possible contexts for negative incorporation. Of these, 452 (87%) are incorporated. Omitting *never*/neg *ever* (where incorporation is almost categorical), the proportion is 76%. In the middle-class interviews the figures are 80% overall and 66% without *never*/neg *ever;* in the lower–class interviews, negative incorporation is 89% overall and 81% without *never*/neg *ever*. Gunnel Tottie points out (personal communication) that in her material there is much more negative incorporation in the written form than in the spoken form. If negative incorporation turns out to be more characteristic of written language in general, this would be one of the rare cases in the Ayr interviews where the language of the lower-class speakers is closer to the written form; usually, it is the middle-class speakers whose language shows characteristics more frequently found in the written form (cf. Biber 1986).

Since negative incorporation is strongly favored, it is not surprising that there should be examples of multiple negative concord (Labov 1972). They are rarer, however, than might have been expected. In the lower-class interviews there are a total of 20. Examples are given in (8):

(8) a. WL392   they said
         393   there was *never nae* gas there
     b. AS1717   I just *couldnae* get a job *naewhere*
     c. MR634   you just *cannae* dae *nothing* aboot it
     d. HG1416   I'll *no* go to *nae* machine

All the lower-class interviews with the exception of Laidlaw's have at least one example of multiple negative concord. (There is one example in Laidlaw's interview, but it was uttered by her husband.) There are also two examples of *hardly* i.e., with a negative:

(9) a. MR373   I *never hardly* missed the school
     b. HG1173 you *couldnae hardly* see oot of the window of the plane

There is also an unusual use of two negatives by Gemmill:

(10)   HG1262 it's not so easy
         263   keeping weans [children]
         264   that they *don't* annoy somebody

Here the meaning is, "It's not so easy keeping weans from annoying somebody." The low frequency of multiple negative concord suggests that it is not the preferred form for lower-class speakers, and it does not appear to be used for special emphasis or rhetorical effect. There is even an example where the absence of negative incorporation may indicate a desire to avoid the effect of multiple negative concord:

(11)   HG316   you *couldnae* get *any* mair nor two pound

The *nor* in 11 is the comparative preposition corresponding to *than*, but negative incorporation might have given the impression of multiple negative concord:

(12)   *you could get nae mair nor two pound

There are no examples of multiple negative concord in the middle-class interviews, though there are four examples of genuine double negation in the interview with Nicoll:

(13)   DN251/52 I *don't* like *not* working
         259   I *can't* do *nothing*
         282/83   I *can't* afford *not* to work
         1016   I *wasn't not* going to go

These are examples of true double negation, and it is unlikely that Nicoll would have used them if he had felt at all anxious about being thought to have used multiple negative concord.

The frequency with which *never* is used in the lower-class interviews raises the question of whether it has a wider meaning than in the middle-class interviews. Cheshire points out that in addition to the standard meaning "not on any occasion," "in most nonstandard British English dialects, *never* also

occurs with a more restricted meaning, that of 'not on one specific occasion' "
(1982, 67). There are only two clear examples of this, both in a story told by
Lang about failing to buy the cheese he had been sent for. Both examples
occur in quoted direct speech:

(14)   a. WL1548   "Whaur's the cheese?"
             1549   "Nae cheese
             1550   he *never* got the cheese"
      b. WL1570   "Did you see Willie?"
             1571   "No I did not Sandy
             1572   I *never* seen him"

These are similar to the examples cited by Cheshire, but most of the examples
are less clear. In many cases the substitution of a simple negative would
change the meaning little if at all, but they are not obviously restricted to a
single specific occasion. It seems more likely that *never* is used as a slightly
emphatic negative and this would account for its occurence before the auxiliary verb, as shown in (15):

(15)   a. HG1394   oh I *never* would go back with a train again
      b. EL1177   well I *never* had seen pointed nails
      c. WL2006   oh I *never* was a picture fan
      d. MR076   I *never* was oot of that hoose

In this position Sinclair frequently makes it more emphatic by including the
unincorporated form as well:

(16)   a. AS434   I *never ever* was back
             435   working at the farm
      b.   2138   a wee man
          2139   that *never ever* was married
      c.   694   I mean I *never ever* was feart
          695   to work
      d.   013   I *never ever* was up there at all

Out of 53 examples of *never* in Sinclair's interview, 20 (38%) are accompanied by *ever*. The only other speaker who does this is Lang, and he has only
2 examples out of 24. Sinclair's discourse style is characterized by a great deal
of redundancy, and his use of *never ever* is an obvious example of this.

    A clearer example of social differentiation (although the numbers are small)
is the negation of main verb *have*. The middle-class speakers generally use
do-support, as in the examples in (17):

(17)   a. IM490   I *didn't have* the time
      b. JM360   I *didn't have* a bike

      c. NM1462   but I *don't have* the thing
      d. DN1994   I *don't have* the best end of the district

Out of 28 examples, there are 26 instances of *do*-support (93%). The two exceptions are by the same speaker, Gibson.

The lower-class speakers, on the other hand, frequently negate main verb *have* with clitic *−nae,* as illustrated in (18):

(18)   a. WL802   and yet I *hadnae* a certificate
         b. MR672   they *hadnae* a hundred of a membership
         c. AS2087   that *hadnae* a bar in it
         d. HG1714   you've got money noo
            1715   that you *hadnae*

There are only 7 cases (23%) of *do*-support out of 30 possible contexts.

In this section we have seen that negation is one area where there is clear social differentiation. The use of *−nae* as a clitic, *nae* as a quantifier, and *no* as the uncontracted negative operator with the verb are all markers of lower-class speech. The use of multiple negative concord is also restricted to the lower-class group. Both social class groups favor cliticization on the subject rather than on the auxiliary and also show a preference for negative incorporation of indefinites, with the lower-class showing a slightly stronger tendency in each case. The middle-class speakers are much more likely to use *do*-support in the negation of *have.*

Subject–Verb Agreement

Another salient characteristic of the lower-class interviews is lack of agreement between subject and verb; it is one of the characteristics that clearly distinguishes the lower-class interviews from the middle-class ones. To investigate this, I searched the transcripts of the lower-class interviews for third-person subject–verb agreement and tabulated the results for each speaker. All present tense verbs (including present perfect) and all examples of the past tense of *be* were scored, with two exceptions: since the reporting verb *say* can be *says* regardless of person (see below), no instances of *say* were counted; and since the form *come* can be either present or past tense, it was also excluded. Subject–verb agreement was tabulated in three categories, as illustrated in (19):

(19)   I. *Following existential* there
         a. WL324   there *were an ice-cream man*
            325   come to the pit on a Friday
         b. AS096   I mean there *was three streets* in Whitletts

   II. *Following a subject relative pronoun*
      c. MR457   *some of the girls* that *was* in it
      d. WL1030   and I mean there are *employers*
          1031   *that's* harmful to the workers just the same
  III. *All other environments*
      e. MR094   and—eh—my brothers was aw working
      d. AS2483   their hands was tied to a certain extent

Three thousand and twenty-four instances of verb agreement were scored in the lower-class interviews. In 2,819 cases (93%) there was verb agreement, and only in 205 (7%) was there lack of agreement. It is thus clear that the saliency of lack of agreement is not because of its frequency. In looking at the individual categories, Category I contains the highest frequency of nonagreement (38%), compared with (19%) in Category II and only 2% in Category III.

Lack of agreement can be in two forms: (1) singular subject with plural verb or (2) plural subject with singular verb. Of the 205 cases of nonagreement 85 (41%) are of singular subject with plural verb, and 120 (59%) have a plural subject and a singular verb. However, the overwhelming majority of examples of nonagreement of singular subject with plural verb (96%) occur after existential *there*. There is individual variation as shown in Table 5–4. It can be seen from Table 5–4 that for all speakers the possibility of nonagreement is highest after existential *there* and next-highest in relative clauses. The most surprising feature of Table 5–4, however, is the fact that Ritchie shows the greatest propensity to nonagreement of all the lower-class speakers. Since even Sinclair shows a fairly high frequency of nonagreement, it seems reasonable to infer that nonagreement of subject and verb is not salient to the speakers themselves, because Ritchie and Sinclair are the two lower-class speakers who show the greatest apparent adaptation to the interview situation. The examples of nonagreement, particularly those with *there* are found fairly

Table 5.4. Lack of Subject/Verb Agreement (%)

Respondent	After *There*	In Relative Clause	Elsewhere	All
Laidlaw	40 (8/20)	3 (3/13)	2 (5/672)	2 (16/705)
Gemmill	39 (20/51)	3 (1/31)	2 (7/355)	6 (28/437)
Sinclair	20 (19/94)	18 (10/59)	3 (19/668)	6 (48/821)
Lang	47 (47/101)	28 (8/29)	1 (3/523)	9 (58/653)
Rae	45 (13/29)	29 (2/7)	1 (2/161)	9 (17/197)
Ritchie	68 (15/22)	50 (3/6)	6 (11/174)	14 (29/202)
All LC	38 (122/317)	19 (27/145)	2 (47/2,553)	7 (196/3,015)

Table 5.5. Nonagreement in *There* Clauses

Type of Nonagreement	Present	Past
Singular subject/plural verb	0 (0%)	82 (100%)
Plural subject/singular verb	25 (63%)	15 (38%)

evenly distributed throughout the interviews and do not appear to be affected by genre or topic.

Of the 122 examples of nonagreement with *there* 40 (33%) are of a singular verb with plural noun phrase following, and 82 (67%) have a plural verb with a singular noun phrase following. There is, however, an effect of tense on subject–verb agreement in *there* clauses. Of the 317 *there* clauses with verbs 83 are in the present tense (including present perfect), and 234 are in the past tense. (There are also 67 *there* clauses with no finite verb.) There is a slightly greater tendency for lack of agreement in past tense *there* clauses, 41%, compared with 30% in present tense *there* clauses. The striking difference, however, lies in the nature of the nonagreement, as can be seen in Table 5–5. The table shows that there are no examples of *there are* + NP[noun phrase] sing, but it turns out that there are also only three examples of *there are* + NPplur in the lower-class interviews as a whole, so that *there are* is not a frequent item. Approximately two-thirds of the examples of nonagreement consist of *there were* + NPsing.

In addition to the 317 *there* clauses with verbs, there are a further 67 with no tensed verb. Examples are shown in (20):

(20)   a. WL2918   there ø naebody going to force them
       b. WR538   there ø always something on ken in—eh—
       c. HG254   there ø a big tree at the gate
       d. HG1212   and to me there ø no trevel like in an aeroplane

Like nonagreement of subject and verb, there is variation among the speakers, as shown in Table 5–6. As with lack of agreement in *there* clauses, Ritchie has the highest proportion of verbless clauses. Since *there are* is virtually absent from the lower-class interviews, it might be thought that the verbless clauses are the result of the deletion of *are*. It turns out, however, that the majority (85%) of verbless *there* clauses have singular NPs, so it is not obvious that the omission of the verb has a simple phonological conditioning. On the other hand, since *there were* is common with singular NPs, the verbless forms may in fact be the nonpast equivalents and thus may originate from *there are*. Approximately half of the verbless examples of *there* are followed immediately by a negative (e.g., *no, nae, naebody,* etc.); but the

Table 5.6. Proportion of *There*
Clauses without Verb

Respondent	Percentage
Sinclair	9
Lang	12
Laidlaw	20
Gemmill	25
Rae	26
Ritchie	35
All LC	17

verb is not always omitted before a negative, so this does not appear to be the conditioning factor either.

All the lower-class speakers make use of a narrative present tense similar to the "conversational historical present" described by Wolfson (1982) and the "historical present" analyzed by Schiffrin (1981). This form occurs in narratives and is usually formally indistinguishable from other uses of the present tense. There are, however, fairly frequent examples of inflected forms of present tense verbs other than in the third person singular. The most common of these is the reporting verb *I says* (83 examples), which occurs only in narratives but is the normal form for reporting the narrator's own speech. There are only 9 examples of *I said* (=10%) and none of *I say* in narratives. There are also 33 examples of *says I* and only one of *said I* and none of *say I*. In other words, 92% of the examples of first-person-singular reporting verbs are inflected present tense forms. This is similar to the third person singular, where 96% of the reporting verbs are in the present tense. There are also two examples of *they says* and one of *we says*. It is clear that an inflected present tense form is the norm for narrative reporting verbs. (There is no evidence of a functional difference between *I says* and *I said*, such as Johnstone [1987] found.) Neither Schiffrin nor Wolfson mentions this characteristic of the narrative present tense, although examples of *I says* occur in their data. It might be thought that this is simply the use of *says* as an invariant form; but *say* occurs as an uninflected form elsewhere, including phrases such as *as I say*. It is only in its narrative function that present tense *say* is inflected other than in its normal use for the third person singular. (Johnstone [1987] reports that in her Northern/North Midland American English corpus from Indiana only *I says* serves as what she calls an "introducer" of quoted direct speech—never *I say*, although the latter is the normal form elsewhere.) Moreover, it is not only *say* that occurs in this narrative tense. There are 13 examples with the verbs *come, cut, get, go, meet,* and *see.* Some of these are given in (21):

(21) a. EL3645 so I *goes* oot this night
      b. WL1513 so I *sees* this
      c. WL1531 of coorse I *comes* hame
      d. AS2582 when here I *meets* the Captain

Although the number of such examples is small, they are sufficiently varied and consistently used to suggest that this is a genuine narrative present tense. Most examples of narrative present tense are third person singular, so that it is impossible to distinguish them from other uses of the present tense; but the first-person-singular inflected forms are used only where a narrative tense would be appropriate. This is a much more restricted use than Cheshire found in Reading (1978, 1982). All the lower-class speakers have examples of this narrative tense, but the only middle-class speaker who uses it is Nicoll (though 55% of his reporting verbs are in the past tense). There is an interesting contrast between his use of *I says* in a narrative context in (22) and his use of *I say* in a reminiscence in (23):

(22)   DN265   he says
           266   "What are you doing?"
           267   I *says*
           268   "I'm doing nothing"
           269   he says
           270   "Would you like
           271   to work?"
           272   I *says*
           273   "Of course I would like
           274   to work"

In (22) Nicoll is referring to a particular occasion that is part of a narrative that also includes this exchange, and *says* has a specific past time reference. In example (23) he is not talking about a specific occasion but about something that frequently occurs:

(23) DN609   but as I say often to people—New Zealanders
      610   when they come in
      611   "Oh you're New Zealanders
      612   which part North or South?"
      613   "Oh we're from the North"
      614   and I *say*

615   "Oh you're not
616   you North people you're not so hot
617   I like the South"

In (23) Nicoll is not referring to a particular occasion, and the verb *say* does not refer to a specific time but is instead a habitual tense similar to that in line 609 (cf. Johnstone's comment [1987, 51]). Nicoll's interview thus demonstrates the appropriate use of *says* as a narrative tense form, although he more frequently uses a past tense verb to introduce quoted direct speech.

Relative Clauses

Although differences in the use of relative clause markers are much less salient than differences in negation, the differences between the two class groups are fairly "robust." The fullest discussion of relative clauses in Scottish English is to be found in Romaine's work on Middle Scots (Romaine 1980b, 1981, 1982a, 1984). The present section owes a great deal to her work and to her comments on an earlier draft. Romaine examined the use of relative clause markers in Middle Scots and modern Scottish English. She has argued that the relative system in Middle Scots marked relative clauses either with TH-forms (e.g., *that*) or ø and that the introduction of WH-forms (e.g., *who, which*) is an infiltration of a rival system, perhaps a borrowing from a Romance acrolectal adstratum in Middle English that "entered the written language and worked its way down a stylistic continuum ranging from the most to the least complex styles" (1980b, 234). Romaine bases this argument on two points. One is that WH-forms are more frequent at the lower end of the noun phrase accessibility hierarchy (Keenan and Comrie 1977; Keenan 1975) and least frequent in subject position. The second is that WH-forms are more frequent on the formal side of the stylistic continuum and become less frequent toward the informal side. Romaine's arguments, naturally, are based mainly on written materials; but she also refers to samples of spoken American English and Scottish English, although in less detail.

In the Ayr interviews there are 1,003 relative clauses, including both restrictive and nonrestrictive relative clauses. Of the total, 54 are so-called sentential relative clauses (e.g.: IM 147/49 in fact—em—I very nearly went into the colonial service *which I'm very thankful I didn't do*), where the relative clause does not modify an antecedent noun phrase. Sentential relative clauses will be excluded from the following analysis because of their different function. (See Romaine 1982a, 84–87 for a discussion of sentential relatives.) Also excluded are relative clauses introduced by the relative adverbs *where* and *when*, of which there are 51 in the interviews. Romaine chose to include them in her totals because they participate in the alternation with TH and ø, though she admits that there is no general agreement on their status. I decided to exclude them mainly because the fact that the range of antecedents is more narrowly constrained than that of the other relative markers seems to affect their distribution. The remaining relatives are classified as either restrictive relative clauses or nonrestrictive relative clauses, though it must be stressed that it is not always easy to decide whether a given instance is restrictive or nonrestrictive (see Romaine 1982a, 81–83; Quirk et al. 1972, 859). Even intonation is not always the clear guide it is frequently said to be, partly because pauses and other hesitation phenomena may disrupt the tone group.

Romaine classifies the 6,062 clauses in her data according to the occurrence of the relative markers in nine syntactic positions: subject, direct object, possessive, indirect object, predicate nominal, temporal, locative, preposition-stranded, and preposition-shifted. Because of the smaller sample and different nature of the Ayr interviews (e.g., there are no relativized genitives), only three categories will be used: subject, direct object, and other syntactic positions. Of the 949 relative clauses in the Ayr interviews, 488 (51%) have the relativized noun phrase in subject position, 298 (31%) are direct objects, and the remaining 163 (17%) are in other syntactic positions. (If the relative clauses introduced by *where* and *when* had been included, the proportions would have been 49% subjects, 30% direct objects, and 21% other positions.) For restrictive relative clauses only, the figures are 401 subjects (48%), 282 direct objects (34%), and 158 in other positions (19%). This is similar to the figures for restrictive relatives only, in Romaine's Middle Scots sample: 65% subjects, 22% direct objects, and 13% in other positions. Romaine's figures for nonrestrictive relatives (57% subjects, 15% direct objects, and 33% in other positions) and for all relatives (61% subjects, 10% direct objects, and 24% in other positions) show a different order of frequency; but this is probably because of the stylistic difference between spoken language and written language in the frequency and use of nonrestrictive relatives.

There is, however, an interesting difference between the Ayr sample and Romaine's Middle Scots sample in the use of WH-forms. Romaine's percentages of WH-forms in restrictive relatives only, are subjects 14%, direct objects 14%, oblique cases 51%, and genitives 75%. In the Ayr interviews, 48% of the subject relative markers, 10% of the direct object relative markers, and 9% of the relative markers in other syntactic positions are WH-forms. (This latter figure would be increased to 30% if relative clauses introduced by *where* and *when* were included.) For restrictive relatives only, the figures are subjects 37%, direct objects 5%, and in other positions 5%. The major difference is in the frequency of WH-forms in subject position.

Table 5.7. All Relative Clauses

Type of Antecedent	Subject	Direct Object	Other	Total
Human	295 (88%)	27 (8%)	12 (4%)	334
Nonhuman	193 (31%)	271 (44%)	151 (25%)	615
Total	488	298	163	949

Other factors, however, affect the distribution of WH-forms. Rosenbaum (1979) claims that there is a tendency for relative clauses with a human antecedent to be subjects and for relative clauses with a nonhuman antecedent to be direct objects. This is confirmed by the Ayr figures, as can be seen in Tables 5–7 and 5–8. Another way of looking at this distribution is shown in Tables 5–9 and 5–10. This bias in favor of human antecedents has an effect on restrictive relative clauses but does not explain the tendency for WH-forms to occur in subject position in nonrestrictive relative clauses, as can be seen in Table 5–11. Although 61% of WH-forms (171/279) have human antecedents, the proportion is much greater for restrictive relative clauses 73% (124/171) than for nonrestrictive relative clauses 44% (47/108). Romaine's Middle Scots sample shows no differentiation of this kind: 52% of the WH-forms in restrictive relative clauses have animate antecedents, and the proportion in nonrestrictive relative clauses is 53%. (Romaine uses the classification animate/inanimate rather than human/nonhuman, but this is unlikely to affect the comparison greatly.)

In Ayr there are also obvious social class differences in the use of relative clauses. The first is that the middle-class speakers are much more likely than the lower-class speakers to use nonrestrictive relative clauses. In the middle-class interviews 20% (84/423) of the relative clauses are nonrestrictive, compared with only 5% (24/526) in the lower-class interviews. This difference does not greatly affect the distribution of subject, direct object, and other relatives, as can be seen in Tables 5–12 and 5–13. Nor is there a great difference in the distribution of WH-forms, as can be seen in Table 5–14. Noteworthy here is that the proportion of WH-forms is much greater in the

Table 5.8. Restrictive Relative Clauses Only

Type of Antecedent	Subject	Direct Object	Other	Total
Human	254 (89%)	21 (7%)	12 (4%)	287
Nonhuman	147 (27%)	261 (47%)	146 (26%)	554
Total	405	282	158	841

Table 5.9. All Relative Clauses

Position	Human Antecedent	Nonhuman Antecedent	Total
Subject	291 (60%)	193 (40%)	488
Direct object	27 (9%)	271 (91%)	298
Other	12 (7%)	151 (93%)	163
Total	334	615	949

middle-class interviews than in the lower-class ones. In the middle-class interviews 53% (223/423) relative clauses have WH-forms, compared with only 10% (55/526) in the lower-class interviews. The difference is even greater if only restrictive relative clauses are considered: the percentage of WH-forms is 41% (140/339) for the middle-class group and 6% (31/502) for the lower-class group. The figures for the lower-class group are skewed, however, by the interview with Sinclair, who provides 26 of the 31 WH-forms in restrictive relative clauses in the lower-class interviews. If Sinclair's figures are omitted, the percentage of WH-forms used by lower-class speakers in restrictive relative clauses is only 1.6% (5/299). The lack of WH-forms in restrictive relative clauses in the lower-class interviews supports Romaine's view that WH-forms are an intrusion into Scots. Their greater frequency in nonrestrictive relative clauses is consistent with Rydén's finding that WH-forms in early-sixteenth-century English are predominantly in nonrestrictive clauses and his claim that WH-forms enter the language first in nonrestrictive relative clauses (1983, 132). This view is also consistent with Romaine's figures for Middle Scots.

It is here, however, that the decision to omit the relative clauses introduced by *where* and *when* becomes crucial. Of the 51 clauses of this type, 20 occur in the middle-class interviews and 31 in the lower-class interviews. This would almost double the number of WH-forms in restrictive clauses in the lower-class interviews and more than triple the number in the interviews other than Sinclair's. If Romaine's decision to include relative clauses with *where*

Table 5.10. Restrictive Relative Clauses Only

Position	Human Antecedent	Nonhuman Antecedent	Total
Subject	254 (63%)	147 (37%)	401
Direct object	21 (7%)	261 (93%)	282
Other	12 (8%)	146 (92%)	158
Total	287	554	841

Table 5.11. Distribution of WH-Forms

Type of Relative	Subject		Direct Object		Other		Total		
	Hum.	Nonhum.	Hum.	Nonhum.	Hum.	Nonhum.	Hum.	Nonhum.	All
Restrictive	121	29	2	11	1	7	124	47	171
Nonrestrictive	41	46	6	10	0	5	47	61	108
All	162	75	8	21	1	12	171	108	279

Table 5.12. Relative Clauses in Middle-Class Interviews

Type of Clause	Subject	Direct Object	Other
Nonrestrictive (n = 84)	65 (77%)	16 (19%)	3 (4%)
Restrictive (n = 339)	180 (53%)	105 (31%)	54 (16%)
All (n = 423)	245 (58%)	121 (29%)	57 (13%)

and *when* is the right one, these figures would support her view that WH-forms "entered the language in the most complex and least frequently relativised syntactic positions in the case hierarchy" (1980b, 233). If the *where/when* relatives are included, however, there is a puzzling contrast in the role they play in the middle-class interviews compared with the lower-class interviews. *Where/when* relatives would account for only 8% (20/243) of the WH-forms in the middle-class interviews but for 36% (31/86) of the WH-forms in the lower-class interviews. Since relative clauses introduced by *where* and *when* bear a strong superficial resemblance to adverbial clauses, it is perhaps interesting that adverbial clauses of time and place are more frequent in the lower-class interviews (6.9 per thousand words) than in the middle-class interviews (5.2 per thousand words). This is consistent with the higher proportion of *where/when* relatives in the lower-class interviews and very different from the proportion of other WH-relatives.

There is, however, one category of WH–relative pronouns that Romaine does not deal with, namely, relative pronouns without antecedent—or what Quirk and his colleagues call "nominal relative clauses" (1985, 1056–61)—of the type illustrated in (24):

(24) a. WG325   and described
          326   *what* was happening
      b. WL152   I developed—runs
          153   *what* they caw'd runs
      c. AS1555 but it's ridiculous—eh—
         1556 *what* they can get away with noo
      d. WR433   that's
         434   *who* I was raised with ken

Table 5.13. Relative Clauses in Lower-Class Interviews

Type of Clause	Subject	Direct Object	Other
Nonrestrictive (n = 24)	22 (92%)	0 (0%)	2 (8%)
Restrictive (n = 502)	221 (44%)	177 (35%)	104 (21%)
All (n = 526)	243 (46%)	177 (34%)	106 (20%)

Table 5.14. Distribution of WH-Forms

Type of Clause/Class	Subject	Direct Object	Other
Restrictive			
Middle (n = 140)	119 (85%)	13 (9%)	8 (6%)
Lower (n = 31)	31 (100%)	0 (9%)	0 (0%)
Nonrestrictive			
Middle (n = 83)	64 (77%)	16 (19%)	3 (4%)
Lower (n = 24)	22 (92%)	0 (0%)	2 (8%)
All			
Middle (n = 223)	183 (82%)	29 (13%)	11 (5%)
Lower (n = 55)	53 (96%)	0 (0%)	2 (4%)

The predominant form is *what,* which accounts for 91% (269/294); but there are 16 examples of *whatever,* 4 of *who,* 3 of *whoever,* and 2 of *where.* These forms are almost three times more frequent in the lower-class interviews. There are approximately 98 cases per thousand subordinate clauses in the lower-class interviews, compared with only 35 per thousand in the middle-class interviews. There is no difference in their distribution. In both social class groups these forms occur most frequently as direct objects, as can be seen in Table 5–15. The relationship between these relative pronouns without antecedent and other relative pronouns is unclear, but it is obvious that the lower-class group displays a willingness to use these WH-forms that is very different from their use of *who/which* though it is not very different from their use of the relative adverbs *where* and *when.* Most noteworthy, perhaps, is the frequency with which these pronouns occur as direct objects, since (as was shown in Table 5–14) *who* and *which* are never used in this position by the lower-class speakers.

Finally, there is a significant difference between the two groups of speakers in their use of ø as the relative marker in subject position, as shown in (25):

(25)   a. NM235   and there was a select number of us
            236   ø became members
     b. WL2054   of coorse nooadays it's every hoose
          2055   ø has got a television sort of thing
     c. WL2341   aye I've a girl
          2342   ø works in Slough—in London

Zero subject markers occur only in restrictive relative clauses. In the middle-class interviews the proportion is less than 3% (5/180), while in the lower-class interviews 24% (53/221) of the subjects are zero. Most of the zero

Table 5.15. Distribution of Relative Pronouns
without Antecedent

Class	Subject	Direct Object	Other
Middle (n = 69)	14 (22%)	44 (71%)	4 (7%)
Lower (n = 223)	57 (26%)	157 (70%)	9 (4%)

subject markers (74%) occur after a clause containing existential *there,* as in
(25a). Another 19% occur in cleft sentences, as in (25b). There are three
examples following a clause with main verb *have,* as in (25c). Zero subject
markers are found in all three contexts in both social class groups. Lower-
class speakers also use more zero relative markers in direct object position
(lower class 71%, middle class 51%) and in other positions (lower class 56%,
middle class 52%), but the differences are less strongly marked than in subject
position.

To sum up, there are obvious social class differences in the use of relative
clauses. Lower-class speakers use nonrestrictive relative clauses less frequently, and *who/which* as the relative marker much less frequently. They use
relative pronouns without an antecedent more frequently and a zero relative
marker in subject position more frequently. In addition, lower-class speakers
use restrictive relative clauses after personal pronouns (e.g.: MR089 *them* that
had wee families), a usage that will be discussed further in the next chapter.

I have shown that there are major differences between the two social class
groups in their use of negation, subject–verb concord, and relative clauses.
These differences are fairly "robust" in the sense that they are not affected by
style or genre. The differences in negation and subject–verb concord are
much more salient and, like phonological features, are susceptible to mimicry.

6

Other Morphological Variables

Other aspects of language show "robust" evidence of social class differences. The present chapter will deal with determiners and pronouns.

The Definite Article

As Wilson points out, *the* is occasionally used in place of a possessive pronoun with reference to close family relatives (1923, 54). There are only a few examples of this from lower-class speakers:

(1) HG325 but at that time *the wife* was living [i.e., the speaker's own wife]

915 *the lassie* wasnae mairrit at this time [i.e., the speaker's own daughter]

AS3518 "You forgot *the boy*" [i.e., the addressee's son]

3520 "No" he says

3521 "I didnae forget *the boy*" [i.e. his own son]

Such examples occur only in the lower-class interviews, but even there they are the exception. Possessive pronouns are much more frequent and are always used with *father* and *mother*.

The definite article is also used:

a. in certain time expressions: *in the day's paper* (= "today's"), *the night* (= "tonight"), *the noo* (= "nowadays"), *the morn* (= "tomorrow"), *at the weekends, at the harvest time, I went to work on the Sunday night* (= "every Sunday night"), *I wasnae fourteen until the August*. This use is found in both middle-class and lower-class interviews.

b. with certain institutions: *at the school, to the church, to the hospital.*
Both middle-class and lower-class speakers use the definite article in
this category, but there is much variation and it is often omitted even
by lower-class speakers.

c. with tasks: *I went to the rope-splicing, I was at the gardening, we
went to the milk* (i.e., delivering milk).

d. with certain expressions of quantity: *he paid the half of what you
were getting, the maist of the farmers kent, when the one supporter
run on the field.*

e. with games and pastimes: *the beds* (i.e., "hopscotch"), *the peeries*
(i.e., "whipping tops"), *at the football, going to the bingo.*

f. in certain expressions of direction: *when he was coming back the
way, up the stair(s), down the stair, they're shoved to the side.*

g. in certain generic expressions: *well the thing aboot the holidays is,
that's the tractors for you, the folk kent to get doon, what they caw'd
the buttermilk.*

All the uses in c–g come from lower-class interviews.

Possessive Pronouns

Possessive pronouns are sometimes used by lower-class speakers in place of
another determiner (including zero): *we were in for wer lunch, for that pool
you could buy your pint of beer your stout or a strong ale or that, to get her
wee blether, by the time you got up and got your breakfast, that spent her
couple of oors up there, you had your singsong, they finally gave us oor
notice.* For the first-person-plural possessive pronoun, two forms are found:
wer and *oor.* Wilson does not give *wer* (1923, 55–57), though he lists it in the
glossary at the end of the book. Grant and Dixon give *wer/wur* as the normal
form for Ayrshire (1921, 96). There seems to be no difference except that *wer*
is never stressed. It is therefore likely to be used when the stress is clearly on
an adjacent item as in (2):

(2)　EL1356　you saw
　　　1357　how we can all sit
　　　1358　and talk *wer ain* tongue *wer ain* way among werselves

There are no examples of *oorselves,* the only form given by Wilson for the
first-person-plural reflexive pronoun.

Demonstratives

One of the more salient markers of social class differences is the use of *thae* and *them* by lower-class speakers where middle-class speakers usually use *those*. In the lower-class interviews there are 21 instances, of which 20 are used by three speakers (Gemmill, Ritchie, and Lang). Rae and Sinclair do not use *thae* at all; but they are responsible for the 12 examples of *them*, 10 of them in Rae's interview. This suggests that *them* may be replacing the older term *thae*. There is only one example of *those*. In the middle-class interviews there is a total of 17 examples of *those*. Wilson remarks, "It is curious to note how commonly a Scotchman, speaking English, uses "these" for 'those,' e.g., 'in these days,' meaning in the distant past where an Englishman would say 'in those days' " (1923, 54). In the lower-class interviews there was one example of *in those days*, 15 of *in these days*, and 21 *in them/thae days*. In the middle-class interviews there were 13 examples of *in those days* and 4 examples of *in these days*.

There is a curious imbalance in the distribution of proximal demonstratives (*this, these*) and distal demonstratives (*that, those, thae, them*) with relation to number. In the total sample there are 510 demonstrative determiners, roughly equally divided between proximal (266, or 52%) and distal (244, or 48%). Of these 510 demonstratives 334 (65%) are in the singular and 176 (35%) in the plural. In the singular, however, 209 (63%) are distal (i.e., *that*), and 125 (37%) proximal (i.e., *this*). In the plural, only 35 (20%) are distal (i.e., *those, thae, them*), and 141 (80%) are proximal (i.e., *these*). Since these are the figures for the total sample, a possible explanation might have been that this imbalance was a consequence of a lower-class tendency to use *these* in place of *those*. It turns out, however, that the proportion of *these* is the same (80%) for both groups.

As with other aspects of language, there are individual differences in the frequency of use. The overall frequency of demonstrative determiners is 3.89 per thousand words, with the lower-class speakers averaging 4.0 per thousand words and the middle-class speakers averaging 3.7 per thousand words. Individual frequencies vary from Gemmill's 1.2 per thousand words to Lang's 9.5 per thousand words, so that frequency of use is not obviously linked to social class.

There are only three examples of *yon* and two of *thon*, as in (3):

(3)   a. WL1754   there were *yon* kind of running boards at the side
                    of it
        b. MR415   I hated *thon* big gate
            416   clanking at the back of me

Pronouns

There are several uses of pronouns that deserve attention. One is the second person plural *youse,* which occurs sporadically in the lower-class interviews. A possible explanation for the rarity of this form is the interview situation where the addressee is the interviewer and a plural form would be inappropriate. There are only 14 examples (Laidlaw 7, Sinclair 4, Lang 3); and half of them are in quoted direct speech. In all cases it is clear that *youse* is a plural form, as in (4):

(4)   a. WL117   and there were maybe four of *youse*
     b. AS1413   whatever you earned was for the pair of *youse*

Interestingly, Laidlaw quotes a teacher as using *youse:*

(5)   EL520   "Annie Laurie would dee
            521   she would die in her box" he says
            522   "if she heard *youse*
            523   there nane of *youse* can sing"

She also quotes herself as using *youse,* though in alternation with *you* in the same reference:

(6)   EL559   I says
            560   "*You* can plunk it
            561   if *you* like
            562   he'll get *youse*
            563   when *youse* are back"

There is no indication in this example that *you* is singular in contrast to *youse,* since both forms are used with the same reference. Wilson does not mention *youse,* and it is probably a comparatively recent import from Northern Ireland via Glasgow. Neither Gemmill nor Ritchie, the two oldest speakers, uses it, though Laidlaw (who is more or less the same age as Ritchie) does. The difference is probably in the nature of the narratives Laidlaw tells.

The middle-class speakers, on the other hand, are the only ones who use the impersonal pronoun *one.* There are 31 examples in the middle-class interviews. But the figure is misleading because only three speakers use this form: Gibson has 21 examples, Menzies 9, and Muir 1. Sometimes, it is clearly a distancing form of *I:*

(7)   a. NM1726   but I don't know
           1727   if *one* was conscious of them—eh—
           1728   being snooty

　　　b. NM437　　*one* doesn't remember it
　　　　　438　　as being very strict
　　　c. WG146　　it was the prestige
　　　　　147　　*one* had

Occasionally, there is an interesting succession of pronouns:

　(8)　a. WG302　　when *I* was a kid
　　　　　303　　*you* played around the beach an awful lot
　　　　　304　　and in the summer time *one* was in a swimsuit
　　　　　305　　oh—seemingly all summer
　　　b. NM1096　　*one* should have taken these chances *I* think
　　　　　1097　　the older that *you* become
　　　　　1098　　*you* realize
　　　　　1099　　that they don't come often

Similar pronoun switching occurs in the lower-class interviews, though not with *one*. Sometimes the effect is somewhat disconcerting, as in the following examples:

　(9)　a. WL648　　so *I* sat down with my piece
　　　　　649　　and aw of a sudden *you* hear this "Psssst!"
　　　　　650　　and the air knocked *your* bunnet aff
　　　b. AS2558　　and this morning *I* come off a watch
　　　　　2559　　and goes to wash *yourself*
　　　c. AS3158　　and *I* think just for peace
　　　　　3159　　*my* faither would just gie *you* a licking or two

Like *youse* for the lower-class speakers, *one* is salient as a class marker, though its use seems to be largely a matter of individual style and cannot be said to be a characteristic of the middle-class interviews as a whole.

　Wilson says that the indefinite pronoun *one* is usually expressed in Ayrshire speech by *a buddie* (1923, 58–59). There is only one example of this, and it occurs in Gemmill's interview in what is almost a formulaic expression:

　(10　HG1641　I don't think
　　　　1642　it did *a body* ony herm

The word *body* also occurs occasionally as a separated form with quantifiers:

　(11)　a. WL407　　if it had been *any* other *body*
　　　　　408　　that had telt me
　　　b. EL748　　and *some* other *body*
　　　c. WL2494　　and there's *nae* other *body* using the bath

Table 6.1. Indefinite Pronouns

Class	Somebody	Someone	% Somebody
Middle	45	3	94
Lower	31	0	100
	Anybody	Anyone	% Anybody
Middle	14	0	100
Lower	11	0	100
	Nobody	No One	% Nobody
Middle	6	0	100
Lower	20*	1†	95

*8 of the lower-class forms are *naebody*
†*no yin* in HG744 and there no yin up

In both sets of interviews, there is a strong preference for indefinite pronouns in *–body* over those in *–one,* as can be seen in Table 6–1. *Yin* occurs only in the lower-class interviews for *one;* but the form *wan* also occurs, though less frequently. The figures for the individual speakers are given in Table 6–2. *Yin* and *wan* are not identical in their use:

a. Only *yin* occurs with adjectives: *the auld yin, the young yins, wee yin, a bad yin, a big yin, a widden yin, the heid yin.*
b. Only *yin* occurs with determiners other than *the* alone: *this yin, that yin, the other yin, another yin, the only yin, the next yin, the last yin, no yin, a yin of them.*

Table 6.2. One and Its Variants

Respondent	Yin	Wan	One	% One
Gemmill	22	8	13	30
Rae	21	4	15	38
Lang	17	1	49	73
Ritchie	3	2	17	77
Laidlaw	7	1	62	89
Sinclair	5	0	88	95
All LC	75	16	244	73

c. Only *yin* occurs in the plural: *they were all yins with big faimlies, there were just odd yins got this, aw the auld yins took their caps off then, thon open yins, some of the young yins don't.*

d. *Yin* occurs in the idiomatic phrase *to be yin for: I wasnae yin for the billiard hall, I was never yin for going into that.*

e. Both *yin* and *wan* occur as the head of a relative clause: *she was the only yin that was up, the yin that we were getting, the wan that's going that road, the wan day we went to Brighton.*

f. Except as the head of a relative clause, *yin* does not occur with the definite article alone, but *wan* does: *I got her over on tae the wan, during the wan of the miners' strikes.*

g. Both *yin* and *wan* occur with *thing: that's yin thing, that was wan thing aboot the older days.*

h. There is only one occurrence of *wan* (and none of *yin*) directly before a noun other than *thing: wan depreciation I dae see.*

i. Both *yin* and *wan* occur as numerals: *twenty-yin, twenty-wan, a hundred tae yin, there were ten chances to yin, I'm going to wan next year, get to a ticket office and get yin.*

The availability of the two forms allows them to be used in contrast, as in the examples in (12):

(12)   a. HG283   there were two blokes
              284   *wan* had a book
              285   and the other *yin* he had money
              286   and this *yin* [i.e., the former] would say . . .
       b. HG1574   the *wan* that's going that road
             1575   he keeps aboot this oot fae the pavement
             1576   the other *yins* they're going this way
       c. MR301   the *wan* went by
              302   and another *yin* came in

There is a parallel contrast to this in the interviews with Gemmill and Rae. *Wunst* is used as a conjunction and *yince* as an adverb, as in the examples in (13):

(13)   a. HG1265   but *wunst* they were auld
       b. HG1500   *wunst* the money wasnae working with them
       c. HG1393   I went with the train *yince*
       d. WR510   you got *yince* a week

Reflexive Pronouns

In the lower-class interviews, there are 41 examples of reflexive pronouns, enough to provide the full paradigm:

(14)   mysel(f)
yoursel(f)
hersel(f)
hissel(f)
wersel(f), werselves
theirsel(f), theirselves

Final [f] is variably deleted. There is one example of *themselves* in the lower-class interviews; and it comes, predictably, in the interview with Sinclair, who, however, follows it almost immediately with *theirself*. In the middle-class interviews, *himself, ourselves,* and *themselves* occur in place of their equivalents in (14). In the lower-class interviews, there are four examples of a reflexive pronoun without a coreferential noun phrase in the same clause:

(15)   a. AS1087   in fact *myself* and the captain of the team
            1092   . . . were the two men
            1093   that went in
   b. AS2468   there was *myself* and another fellow fae Ayr
   c. EL1992   it was folk like *werselves*
   d. HG187   there was only *mysel* aboot this place

There are also two examples of coreferential pronouns within the same scope that are not reflexive:

(16)   a. HG1709   I've seen *me*
            1710   working holidays and weekends
   b. AS1265   well that was you
            1266   giving *you* a circulation for air you see

There are no examples similar to those in Items 15 and 16 in the middle-class interviews.

The examples so far have illustrated certain differences between the two social class groups in the form of pronouns used. Some forms, such as *thae, youse, yin, wan,* and some of the reflexive pronouns, are salient markers of lower-class speech even though their overall frequency is not great. A slightly less salient but equally marked characteristic of middle-class speech is the use of the impersonal pronoun *one.*

Table 6.3. Frequency of Personal Pronouns

	Middle-Class		Lower-Class	
Person	n	/1,000 wds.	n	/1,000 wds.
1st person	3,471	68.2	3,621	53.2
2d person	833	16.4	1,711	25.1
3d person	2,354	46.2	4,913	72.1
Total	6,658	130.8	10,245	150.4

There are also clear differences between the two social class groups in their use of personal pronouns. The lower-class speakers not only use pronouns more frequently than middle-class speakers, they use them in different ways. The figures for the two social class groups are shown in Table 6–3. These figures do not include the use of *you* and *I* in the discourse markers *you know*, *you see*, and *I mean*, since such items are in a sense outside of the syntactic structure. (Since such discourse markers are more frequent in the lower-class interviews than in the middle-class ones, as will be shown in Chapter 11, their inclusion would generally increase the difference between the two social class groups.) It can be seen from Table 6–3 that personal pronouns in general are more common in the lower-class interviews but that first-person pronouns are more frequent in the middle-class interviews, accounting for almost half of the pronouns, compared to one-third of the lower-class personal pronouns. Conversely, the lower-class speakers make much more frequent use of second- and third-person pronouns. In Biber's factor analysis of the differences between spoken and written English, the use of first- and second-person pronouns is one of the features of the "dimension which characterizes texts produced under conditions of high personal involvement" (1986, 394), whereas third-person pronouns are part of the factor that "distinguishes texts with a primary narrative emphasis" (p. 396). Since a greater proportion (29%) of the lower-class interviews consists of narrative (compared with 22% in the middle-class interviews), this may account for the higher frequency of third-person pronouns in the lower-class interviews. (Biber, however, treats the third-person inanimate pronoun *it* separately from the other third-person pronouns; and it appears as part of the first dimension of high personal involvement.) It is *how* pronouns are used, however, that most clearly distinguishes the lower-class speakers from the middle-class speakers. In particular, the lower-class speakers place pronouns in a prominent position in the discourse, as can be seen in several different constructions. The first is illustrated in (17):

(17) a. WL579   so that was *me* on the rope-splicing
       b. WL791   and this was *him* landed with a broken leg
       c. WL1474   that was *him* idle
       d. EL394   and that was *you* shut in the house for a week
       e. EL3251   so that was *us*
       f. AS2248   and that was *it* you see
       g. HG583   and that was *you* maybe slaving until eicht or
                       nine at nicht

In this construction, which I will refer to as *demonstrative focusing,* a demonstrative pronoun is followed by some form of the verb *be* and the object form of a pronoun, usually followed by a complement, which can be a present participle (17g), a past participle (17b, 17d), an adjective (17c), or an adverbial phrase (17a). Although examples are not frequent (23 in the lower-class interviews as a whole), all the lower-class speakers have at least one example, and there are no examples in the middle-class interviews. This construction is one of several where pronouns are given a prominent focus, and it occurs either at the climax or the conclusion of a narrative or a reminiscence.

Demonstrative focusing is not totally unlike *it*-clefting, where *it* occurs as a dummy subject with a form of the verb *be* followed by a noun phrase and a relative clause. There are 68 examples of *it*-clefts in the Ayr interviews, 57 (84%) in the lower-class interviews, and 11 (16%) in the middle-class interviews. In the middle-class interviews the focused NP is always a lexical NP; but in the lower-class interviews it can be a pronoun, as shown in (18):

(18) a. WL1411   it was *her*
             1412   that got them
       b. EL660   no it's *me*
             661   that's het [i.e., "it" in a game]
       c. EL3118   it was *him*
           3119   that led the band
       d. MR639   it's *them*
             640   that's running it

Again, such examples are not frequent even in the lower-class interviews; but they are quite salient, and they have the effect of emphasizing pronouns without using emphatic stress.

Another construction in which pronouns are given greater prominence is NP-fronting, similar to Y-movement. There are 40 examples of NP-fronting in the Ayr interviews, 10 in the middle-class interviews and 30 in the lower-class interviews. However, it is only in the lower-class interviews that NP-fronting has the effect of leaving a pronoun in clause-final position, as in the examples in (19):

(19) a. WL1875   Davie Drummond you called *him*
      b. EL1435   Wee Maggie Rose we called *her*
      c. AS2612   the Army and Navy Club they called *it*

Although the pronoun is not stressed in final position, its relative isolation gives it slightly more prominence than it would have had in the unmarked word order. The explanation for this usage probably lies in the development of relative clause markers in Scottish English. As Romaine (1982a) has shown, WH-forms (e.g., *who, which,* etc.) are an innovation in Scottish English; and as was shown in the previous chapter, such forms are very rare among the lower-class speakers in the Ayr sample. Moreover, the possessive WH-form *whose* is totally lacking from the Ayr interviews, so that the construction *whose name was . . .* does not appear to be a possible alternative. One speaker uses a prepositional phrase instead, as shown in (20):

(20)   EL1170   and a wee woman *with the name of old Maggie Jam-*
                 ieson took me to this machine

Although the middle-class speakers do not use *whose* either, a construction with a nonrestrictive relative clause and passive voice provides an alternative, as shown in (21):

(21)   IM079   and I went to Prestwick High School
          080   *which it was called* then

The use of the passive would not be unusual for the lower-class speakers, but the nonrestrictive relative clause would be. It was shown in the previous chapter that the lower-class speakers rarely use nonrestrictive relative clauses, which account for less than 5% of the relative clauses in the lower-class interviews, compared with 20% in the middle-class interviews. The avoidance of WH-forms even where the head is a pronoun can be seen in (22):

(22) a. AS2990   *him*
         2991   *that's* now head of the union thing
      b. MR089   *them*
         090   *that* had wee families

Such personal pronoun antecedents are never found in the middle-class interviews where the alternative would be an indefinite pronoun, as in (23):

(23)   JM819   *the ones*
                 820   who had been working

Once again, the lower-class speakers use personal pronouns in a more prominent position than the middle-class speakers.

The final example is the use of right dislocation, where a noun phrase occurring in clause-final position is coreferential with a pronoun earlier in the same clause. The coreferential pronoun can be either a subject or an object, as shown in (24):

(24) a. WL1373   in fact *he* offered me a job *Mr Cunningham*
      b. WR786   *she* was a very quiet woman *my mother*
      c. EL3786   and we'll play a trick on *her old Mrs McNaught*
      d. HG772   I was asking John
          773   if he ever heard of *it Cabbies Kirk*

When the rightmost NP is itself a pronoun, the coreferential pronoun must be the subject, as in (25):

(25) a. WL1315   *he* was some man *him*
      b. HG859   but *she* was a harer *her*
      c. HG1348   oh *it* was a loss *it*
      d. WL2728   oh *he* was some dog *him*

The effect of repeating the pronoun in clause-final position is made more emphatic by the fact that the final pronoun is in the stronger object form.

In four different constructions—demonstrative focusing, *it*-clefts, NP-preposing, and right dislocation—the lower-class speakers bring personal pronouns into prominence in a way that is not found in the middle-class interviews. This seems to be part of a general strategy for highlighting or expressing intensity, which Labov says is the linguistic feature "at the heart of social and emotional expression" (1984, 43). The strategy is the use of movement rules, which is overwhelmingly more common in the lower-class interviews, as can be seen in Table 6–4. The table shows that movement rules of

Table 6.4. Highlighting Movement Rules

	Middle-Class		Lower-Class	
Type of Rule	n	/1,000 wds.	n	/1,000 wds.
Left dislocation	7	.14	56	.80
Right dislocation	1	.02	37	.53
NP-fronting	10	.20	30	.43
it-clefts	11	.22	57	.82
Demonstrative focusing	0	.00	23	.33
Total	29	.58	203	2.91

this kind are approximately five times more frequent in the lower-class interviews than in the middle-class ones. Does this mean that the middle-class speakers do not express intensity as frequently? To some extent, that is the case; but it is also true (as will shown in more detail in Chapter 9) that the middle-class speakers express intensity differently, through the use of adverbs, as illustrated in (26):

(26)   a. IM107   I found it *extraordinarily* boring
       b. IM292   I got *absolutely* sick of doing nothing
       c. NM1381   oh well Jock it's *awfully* tasty
       d. DN152   a *terribly* crippled bent old woman

The lower-class speakers rarely use adverbs in this function. In fact, adverbs as specific emphatics and intensifiers, specific hedges and down-toners, attitudinal disjuncts, and style disjuncts (see Biber 1986, 411 for examples) are at least three times more frequent in the middle-class interviews (see Chapter 9).

Why do the lower-class speakers highlight personal pronouns in intensification contexts? Although the evidence is too slender to make a strong case, it seems likely that pronouns carry a greater functional load for the lower-class speakers in comparison with the middle-class speakers. This can also be seen in the use of third-person anaphoric pronouns with no antecedent in the immediately preceding linguistic context. Sometimes the examples fall into the category that Prince calls "inferrable entities" (1981, 236), as in (27):

(27)   HG1469   at breakfast the *taxi* was waiting
            1470   streets werenae too busy
            1471   but when *he* came to lights
            1472   that were against *him*
            1473   *he* was up a side street and away

Although there is no antecedent, it is clear that *he* refers to the driver of the taxi, which *has* been mentioned. Du Bois states this relationship as a general principle: "To make a definite reference to an object, it is not necessary for there to be in the previous discourse a reference to the object; it is only necessary for the idea of the object to have been evoked in some way" (1980, 215). The lower-class speakers have apparently a different sense of the frame that can help identify the referent of an anaphoric pronoun. At other times, the reference is to someone present but previously mentioned:

(28)   a. EL804   because when *his* mother and my mother were at
                     school . . . [*his* = "her husband's"—he was present but not then taking an active part in the discussion]

b. CC405   the likes of *their* parents—*they've* bought a house
[*their* = "the children's"—they were playing in
the room where the interview was taking place]

In neither case was the deictic nature of the reference made clear by stress or gesture.

Sometimes the referent must be inferred from the wider context. The next examples from Lang need more extensive quotation to be appreciated:

(29)   1436   and when I sterted in the pit
         1437   it was the boyne
         1438   for washing
         1439   I mean there were three of us going into the hoose there
         1440   you're oot into the boyne in the washhouse
         1441   there were nae such thing as baths
         1442   you see
         1443   what I mean
         1444   these were the olden days
         1445   you'd
         1446   to get into a tub
         1447   and get youself washed oh aye
         1448   oh the baths are a great thing
         1449   oh they're a great thing
         1450   it meant
         1451   if you were coming hame fae the pit
         1452   and you sat doon
         1453   that was you till aboot six o'clock
         1454   afore you got washed
         1455   and it was always the same
         1456   "Right get yourself washed
         1457   before you fall asleep"
         1458   that was always the cry of my mother
         1459   "Get yourself washed
         1460   before you fall asleep"
→   1461   and *he* was a great man
         1462   the bus come at twenty-five minutes to six in the morning
         1463   and he started at seven o'clock
         1464   twenty-five to six
         1465   now see if he didnae get that bus
         1466   that was him
         1467   he wasnae oot

        1468   he needed
        1469   to get the first one in the morning
        1470   going doon the pit
        1471   or that's him slept in
        1472   so if the bus was by at twenty-five minutes to six
        1473   that was him idle
        1474   no way would he go on the next bus
        1475   there were maybe three buses going
        1476   but no he had
        1477   to get the first one

There are many features of Lang's narrative style that will be discussed later, but here the point is the pronoun in line 1461, "and he was a great man." There has been no reference to Lang's father in the previous 20 minutes, but the identification is quite clear. This extract leads straight in to a story about his father, although his father is not actually mentioned until the middle of the story. Yet it is clear that Lang is telling this story *because* it is about his father and supplies further evidence that he was indeed a great man. For this reason, it is very important that the correct identification of the referent in line 1461 be made; but Lang leaves it to the listener to make this inference.

That this is not an isolated example can be seen from another extract later in the same interview. Lang has been describing a visit to the United States and this is the conclusion:

        (30)   2074   we were all roond New York
                2075   and then we flew back
                2076   come back across in the plane
                2077   and we got to Gatwick
                2078   I think
                2079   it was Gatwick aye Gatwick
                2080   and I think
        →   2081   *she* thought
                2082   I was running away fae her at that time
                2083   I don't like the aeroplanes no no no no
                2084   I was ready for the toilet
                2085   when we landed
                2086   I think
                2087   she thought
                2088   I was running away from her

Although there has been no previous reference to his wife, the reference to her in line 2081 is clear.

Implicit reference of this kind and pronoun switching of the kind illustrated in (29) and (30) place a certain burden on the listener, though there is usually little problem in identifying the referent. One example of real-life failure of identification is given by Laidlaw in an anecdote about the community center:

(31)	EL3839	now the other night Thursday night of the meeting
	3840	after the meeting was finished
	3841	he said to me
	3842	"Check these toilets for me Ella
	3843	because I don't like
	3844	going in
	3845	in case there anybody in
	3846	any one with ladies in it"
	3847	so I goes into the toilets [rest rooms]
	3848	I says
→	3849	"Willie *it's* letting in water"
	3850	he says
	3851	"What's letting in water?"
	3852	"Your toilet's flooded"
	3853	"That's thae bloody lassies"
	3854	he was mad
	3855	he says
	3856	"Look"
	3857	every stopper was in one of the wee washhand basins

"Thae bloody lassies" had put the stoppers in the basins and left the taps running, causing the place to be awash.

This example illustrates how even members of the community may have problems with implicit reference. Willie has to ask what the referent is for *it* in line 3849. On the other hand, in the same interview there is a good example of implicit reference in an exchange between Laidlaw and her husband (JL = Laidlaw's husband):

(32)	EL3972	they were on strike during the war
	3973	what was it for?
	3974	what was the argument during the war?
→	JL	oh *he* steys up there
	EL3975	oh Chairlie Gordon
	JL	he said he couldnae dae the tally
	EL3976	that's right

Here Laidlaw is immediately able to identify the pronoun referent in her husband's response even though the description would appear to fit a number of males living in the vicinity.

This is the kind of implicit reference that Bernstein (1971) found characteristic of a "restricted code." Like many of Bernstein's observations it is probably a half-truth. In many ways the lower-class speakers are *more* explicit than the middle-class speakers (e.g., in their use of highlighting movement rules); but they expect their listeners to be paying attention, and among the things that require careful attention are pronouns. By their use of highlighting devices they remind the listener that pronouns are not just anemic copies of other noun phrases but full-blooded linguistic items capable of referring to people and things in the real world.

7

Further Syntactic Variation

The syntactic features that have been examined so far display relatively "robust" social class differences, partly because they are fairly salient and partly because they are not influenced greatly by genre or topic. In fact, they are typical of the kind of syntactic features that have been dealt with in sociolinguistic studies (e.g., copula deletion [Labov 1969]; negative concord [Labov 1972]; *that*-deletion [Kroch and Small 1978]; *ne*-deletion in Montreal French [Sankoff and Vincent 1977]; and *que*-deletion in Montreal French [Sankoff 1980]). In other words, these studies deal with syntactic markers rather than syntactic structures; and such markers share with phonological variables the characteristics of being high in frequency, relatively immune to conscious suppression, integral to larger structures, and easily quantifiable (Labov 1966a, 49). The use of such markers, however, gives little information regarding the distribution or frequency of the syntactic structures in which they occur.

It is hardly surprising that sociolinguists should have paid little attention to syntax (though it is perhaps paradoxical that this should have happened during a period when syntactic questions were the major preoccupation of most generative grammarians). The primary reason was no doubt the methodology of collecting speech samples, which limited the kind of syntactic structures recorded. A second reason was probably the immense amount of time that transcription and classification requires prior to even the most preliminary analysis; a consequence of this is that a considerable investment of time is necessary, with no guarantee that the results will be interesting. A third problem, as Lavandera (1978) and Romaine (1984) pointed out with reference to Weiner and Labov's (1983) study of the passive, is that the semantic equivalence of variants is hard to establish. Fourthly, there is the problem of identifying the nonoccurrence of a variant (Labov 1982). Labov concluded that quantitative work in syntactic structures would probably be "confined to tracing the relative frequency of occurrence in some globally defined section of speech, controlled for length by an independent measure like number of sentences, pages, or hours of speech" (1982, 87). This is the approach taken in the present chapter, but even such an apparently straightforward task runs into problems of identification of units. Even the notion *subordinate clause* is problematic (Hunt 1965; G. Lakoff 1987; Thompson 1984, 1987); and the question is further complicated by the possibility of distinguishing between embedding and hypotaxis (Halliday 1985). The problems are aggravated by the occurrence of false starts and repetitions as well as interpolated remarks in the interview speech, all of which can make it difficult to determine clausal boundaries and hypotactic or paratactic relationships.

I shall look at the distribution of syntactic structures in the two groups of speakers to determine whether there are differences that might be related to social class. Obviously, the frequency of clause types largely depends upon the content of the utterances and the kind of information being communicated. Unfortunately, the analysis in the present chapter is not sufficiently fine-grained to track the distribution in different genres accurately, partly because it is difficult to establish objective criteria for genres other than a distinction between narrative and nonnarrative sections of the interview. The analysis instead concentrates on the overall characterization of the individual interviews, showing the kinds of gross differences that appear between the lower-class and the middle-class interviews.

The clauses with finite verbs fall into three categories: main, coordinate, and subordinate. There is a small but possibly important difference between the middle-class speakers and the lower-class speakers in the proportion and frequency of these clauses, as can be seen in Table 7–1. It can be seen from Table 7–1 that the middle-class speakers use a proportionately higher number of subordinate constructions. The percentages in the final line are equivalent to what Lou LaBrant called the "subordination index" or "subordination ratio" (cited in Hunt 1965, 13). As is to be expected, however, there is considerable variation within each group, as shown in Table 7–2. It can be seen from Table 7–2 that Sinclair's subordination index (35.5%) is more than twice Rae's (15.7%) and almost twice as high as Ritchie's (20.8%). There are a number of possible reasons for this, one of which is that Sinclair's interview is four times longer than Rae's and five times longer than Ritchie's, so that there are more opportunities for varied syntax. But in subordination, as in many other features, Sinclair's interview is an anomaly; and the differences in subordination cannot be attributed solely to the length of the interviews. For example, Lang's subordination index is only 18.4, although his interview is almost as long as Sinclair's. As in many other instances, it is necessary to

Table 7.1. Clause Distribution

Type of Clause	Middle-Class			Lower-Class			All		
	%	/1,000 wds.	n	%	/1,000 wds.	n	%	/1,000 wds.	n
Main	40.1	53.1	2,703	47.3	63.5	4,431	44.3	59.1	7,134
Coordinate	29.5	39.0	1,985	27.7	37.2	2,595	28.4	37.9	4,580
Subordinate	30.4	40.2	2,047	25.0	33.6	2,348	27.3	36.4	4,395

Table 7.2. Clause Distribution: Lower-Class Speakers

Respondent	Main			Coordinate			Subordinate		
	%	/1,000 wds.	n	%	/1,000 wds.	n	%	/1,000 wds.	n
Rae	65.6	87.6	438	18.7	25.0	125	15.7	21.0	105
Lang	50.0	57.7	940	31.6	36.4	594	18.4	21.2	345
Ritchie	57.0	83.4	367	22.2	32.5	143	20.8	30.5	134
Laidlaw	54.7	83.2	1,098	22.8	34.7	458	22.5	34.3	453
Gemmill	41.1	70.0	627	34.6	53.9	528	24.3	37.8	370
Sinclair	36.3	45.3	961	28.2	35.2	747	35.5	44.4	941
All LC	47.3	63.5	4,431	27.7	37.2	2,595	25.0	33.6	2,348

consider the lower-class group without Sinclair, in order to see the typical distribution. Without Sinclair, the subordination index for the lower-class group is only 20.9%. (The overall frequency of subordinate clauses, however, rises to 29.0.)

The middle-class group is represented in Table 7–3. In this group there is also one speaker whose subordination index deviates greatly from those of the rest of the group. MacDougall is by far the best-educated of the sample; and (as the more detailed analysis of subordination will show) some characteristics to his speech distinguish it from that of the other middle-class speakers. Without MacDougall, the subordination index for the middle-class group is 26.8% (and the frequency of subordinate clauses is 36.9). The skewing effect of two individuals, Sinclair and MacDougall, is obvious in this example and would be troubling if the intention were to generalize from the group scores alone as representative of the population as a whole. Instead, this analysis shows that there is in the proportion and frequency of clause types a *continuum* relating to social class and education, in contrast to the more "robust" features (such as certain phonological and morphological variables) that dichotomize the sample. In more "fragile" features, such as the proportion and frequency of subordination, there is overlap between the two classes. The aim of the present analysis is not to claim that either small group of speakers is typical or representative of the social class divisions in the community as a whole but rather to show which linguistic features are sensitive to social class differences.

It is not only the proportion of subordination that is important; there is also the question of the syntactic complexity that results from greater use of subordination. Measuring syntactic complexity is a controversial topic (Macaulay 1986; Broeck 1977) but two examples in the literature are helpful. First, Hunt identified what he called the T-unit or "minimal terminable unit" (1965, 21), which is "the shortest grammatically allowable" sentence into which an extended utterance could be segmented. Hunt was dealing with school-children's written language, and he wanted a unit that would not be dependent on the idiosyncrasies of their punctuation. Hunt rejected the notion of measuring depth of clause subordination, partly because of the rarity of multiple embedding and partly because the coordination of subordinate clauses (parataxis) confounds the measure.

Second, Arena developed a "clause analysis technique for measuring style complexity" that included the notion of "embedding depth" (1982, 147). A major difference between the approaches of Arena and Hunt is that Arena includes nonfinite syntactic units as separate clauses, while Hunt treats nonfinites as part of the clause in which they appear.

Both Hunt and Arena are dealing with written materials, where it is easier

Table 7.3. Clause Distribution: Middle-Class Speakers

Respondent	Main			Coordinate			Subordinate		
	%	/1,000 wds.	n	%	/1,000 wds.	n	%	/1,000 wds.	n
Nicoll	41.3	53.0	811	36.0	46.1	706	22.7	29.2	446
Menzies	50.8	76.9	392	25.0	37.8	193	24.2	36.7	187
MacGregor	37.5	45.6	351	32.9	40.0	308	29.5	35.8	276
Gibson	41.9	57.1	485	26.1	35.5	302	32.0	43.6	371
Muir	45.0	71.1	327	22.6	35.7	164	32.5	51.3	236
MacDougall	28.6	34.4	337	26.4	31.8	312	45.0	54.2	531
All MC	40.1	53.1	2,703	29.5	39.0	1,985	30.4	40.2	2,047

to identify units and their boundaries than in spoken utterances. Nevertheless, their approaches to the problem of characterizing the range of subordination suggested a way of quantifying the differences among the Ayr speakers. Minimal sequences of syntactic units, including nonfinite syntactic units, were tabulated for each speaker. In practice, this means a main or coordinate clause with the subordinate clauses (and embedded structures) that necessarily depend upon it. These *minimal complex syntactic units* (MCSUs) were scored with the main or coordinate clause counting as zero and each dependent subordinate clause as one. Some of the examples are very easy to score, as illustrated in (1) and (2):

(1) AS1059   we were the team
     1060   that was on duty down the pit at the time
     1061   when they decided
     1062   to seal off the section in Kames
     1063   where the explosion had taken place

The main clause is that in line 1059, and the subsequent four lines each contain syntactic units that are dependent on it. The example in (1) received a score of four.

(2) AM481   and I think
     482   it's a shame
     483   that it's not made a bit easier
     484   or at least that graduates with the appropriate disci-
           plines aren't encouraged
     485   to come into medicine
     486   because I think
     487   in this respect British medicine is a wee bit behind
           United States medicine
     488   where—where people do have a scientific training first

The example in (2) received a score of seven. In a number of cases, however, it is less easy to determine the dependency relationships, as can be seen in (3):

(3) AS2680   and an awful lot of people
     2681   that go abroad
     2682   they go with a false impression
     2683   that I mean
     2684   what I had saw abroad in my experience of the mer-
           chant navy
     2685   I mean I knew
     2686   that it was nae different

Although the actual syntax is awkward, it is reasonable to treat this sequence as an MCSU that could be edited to (4).

(4)   a. and an awful lot of people that go abroad
      b. go with a false impression
      c. that (from) what I had saw abroad in my experience in the merchant navy
      d. I knew
      e. was nae different

This would receive a score of four. Most of the examples that are difficult to score are less tortuous than Item 3 and make up less than 10% of the whole. Nevertheless, the figures that follow should be taken as approximate rather than exact, since in almost every interview there were examples that required an arbitrary decision in scoring. The figures in Table 7–4 show the proportion of MCSUs with one, two, three, and more-than-three dependent syntactic units for the lower-class speakers. Also included is an MCSU index calculated by summing the total number of dependent syntactic units for each speaker divided by the number of MCSUs. This is not a depth-of-embedding measure, since many of the dependent syntactic units are at the same level; but it gives a simple way of comparing the scores for each individual. Also included is the longest MCSU found in each speaker's interview. The striking point about Table 7–4 is that the results are so consistent. With the exception of Sinclair, whose atypicality is clearly revealed, the speakers show a very similar pattern, with most subordinate clauses occurring alone with a main or coordinate clause and less than 10% with more than one other subordinate clause. The figures for the middle-class speakers, (again with one exception) show a similar consistency, although longer MCSUs are more frequent, as can be seen in Table 7–5. Again, MacDougall is the exceptional speaker, with twice

Table 7.4. Lower-Class Speakers: Proportion of Minimal Complex Syntactic Units

Respondent	Length (%)				Longest MCSU	MCSU Index
	1	2	3	4+		
Rae	83	12	2	3	5	1.28
Ritchie	76	18	5	1	4	1.29
Lang	76	19	3	2	5	1.31
Laidlaw	70	22	7	1	5	1.40
Gemmill	72	20	5	4	6	1.40
Sinclair	60	22	11	7	10	1.72

Table 7.5. Middle-Class Speakers: Proportion of
Minimal Complex Syntactic Units

Respondent	Length (%)				Longest MCSU	MCSU Index
	1	2	3	4+		
Menzies	72	21	5	2	7	1.39
Muir	64	25	7	4	7	1.57
Nicoll	63	24	8	5	7	1.57
MacGregor	59	26	8	7	7	1.69
Gibson	57	26	9	8	6	1.72
MacDougall	44	26	15	15	12	2.44

as many MCSUs of three or more dependent syntactic units as any of the other middle-class speakers. Even if the two exceptional cases, Sinclair and Mac-Dougall, are omitted, there is a clear difference between the two social class groups, as can be seen in Table 7–6. It can be seen from Table 7–6 that there is a tendency for the lower-class speakers to use shorter MCSUs, though Sinclair in fact uses longer MCSUs than any of the middle-class speakers, with the exception of MacDougall. This shows that the complexity of subordination is not strongly determined by social class or education, since Sinclair is certainly less well educated than all the middle-class speakers. The syntactic complexity of Sinclair's remarks is consistent with the somewhat pedantic and careful phrasing of much of his interview. On the other hand, Mac-Dougall's language reflects not only his greater education but the elaborated style of academic discourse, as illustrated in (5):

(5) AM280   but—but I found it very difficult
    281   to get a job
    282   or get into a type of situation
    283   where I'll be able

Table 7.6. Proportion of MCSUs by
Social Class Group

Class	Length (%)			
	1	2	3	4+
All middle-class	58	25	9	8
All lower-class	69	21	7	4
Middle-class w/o MacDougall	62	24	8	6
Lower-class w/o Sinclair	74	20	5	2

284   to run a research program on that basis
285   because the—the research departments and the—the
        research direction tended
286   to lie in the—the hands of medically qualified people
287   who often didn't
288   or who often weren't *au fait* with physical-chemical
        techniques
289   so that it meant
290   that you had
291   to convince them
292   that something was worth
293   doing
294   before you could get the equipment

Regardless of the method of analysis, this kind of syntax, with its intricate and involved hypotaxis, is more complex than that found in any of the other interviews; and the hesitations and repetitions probably reflect the kind of internal processing that is going on (Goldman-Eisler 1968). It also illustrates the kind of "grammatical intricacy" that Halliday (1987) claims is characteristic of spoken language.

The average number of words per finite clause was also calculated, and the results for both social class groups are shown in Table 7–7. The most surprising feature of Table 7–7 is the similarity of the figures both between and within groups. The total number of clauses tabulated is over 16,000—9,374 in the lower-class interviews and 6,735 in the middle-class interviews—yet the average length varies little. Both for the groups and for individual speakers, coordinate clauses are approximately one word longer—presumably, the conjunction. Subordinate clauses are more variable—not surprising, since the conjunction is often omitted. The consistency of the figures in Table 7–7 supports Bever's (1972) view of the clause as a perceptual unit, and the length is consistent with Miller's (1956) account of the magical number seven. Chafe claims that intonation units in English "have a modal length of five words" (1986b, 145); and Pawley and Syder (1977) argue that the maximum length for fluent spoken clauses is seven to ten lexical units. It may be worth pointing out that Laidlaw and Lang, who are very fluent speakers, have a somewhat shorter average clause length than three of the more hesitant speakers (Sinclair, MacGregor, and MacDougall); but it would take a more detailed comparison to establish a correlation between clause length and fluency.

Since negation adds to the semantic complexity of a clause, it is a plausible hypothesis that negation will be more frequent in main clauses, assuming that in some undefined sense main clauses are simpler than coordinate or subordi-

Table 7.7. Average Length of Finite Clauses in Words

Respondent	Main	Subordinate	Coordinate	All
Lower-class				
Gemmill	5.7	5.4	6.3	5.9
Laidlaw	5.3	5.1	6.0	5.4
Lang	5.5	5.6	6.6	5.8
Rae	6.6	5.9	7.8	6.7
Ritchie	5.7	5.3	6.4	5.8
Sinclair	7.0	7.1	7.1	7.1
All LC	5.8	6.0	6.6	6.2
Middle-class				
Gibson	5.4	6.3	6.7	5.8
MacDougall	6.2	6.9	7.1	6.5
MacGregor	6.9	7.0	7.6	7.2
Menzies	5.5	6.1	7.4	6.1
Muir	5.1	6.0	6.8	5.8
Nicoll	5.4	5.9	6.5	5.9
All MC	5.6	6.4	6.9	6.2
Total	5.8	6.2	6.7	6.2

nate clauses. This prediction is borne out by the figures in Table 7–8. For the group as a whole, negation is twice as frequent in main clauses as in subordinate clauses. The figures for coordinate clauses are misleading, however, as Table 7–9 shows. Negation is much more frequent after the coordinating conjunction *but* than in coordinate clauses introduced by *and*. This is not surprising, since *but* differs from *and* in having an adversative meaning.

Since the use of the passive is often considered a possible indicator of stylistic variation, the total number of occurrences of the passive in each interview was tabulated. The figures for both social class groups are given in Table 7–10. Contrary to what has been asserted elsewhere (e.g., Bernstein

Table 7.8. Finite Clauses with Negation (%)

Type of Clause	Lower-Class	Middle-Class	All
Main	18	18	18
Coordinate	11	11	11
Subordinate	8	10	9

Table 7.9. Coordinate Clauses with Negation %

Type of Clause	Lower-Class	Middle-Class	All
And	6	8	7
But	25	24	25

1971), there is no significant difference between the two social class groups in their overall use of the passive. The overall frequency is, in the lower-class interviews, 4.17 per thousand words and in the middle-class interviews, 3.56 per thousand words. It is true, however, that the proportion of *get*-passives is much greater in the lower-class interviews (20.1%, compared with 4.4% in the middle-class interviews). The frequency of *be*-passives alone is almost identical (lower-class 3.3, middle-class 3.4 per thousand words). As can be seen in Table 7–10, *get*-passives almost categorically have animate subjects. It is also not the case that passives are absent from narratives; in the lower-class interviews 26% of the passives occur in narratives, in the middle-class interviews 15% of the passives are in narratives.

Table 7.10. Number of Verbs in the Passive

	Be		Get			
Respondent	Animate	Inanimate	Animate	Inanimate	Total	/1,000 wds.
Lower-class						
Rae	4	6	3	0	13	2.6
Gemmill	10	6	11	0	27	2.8
Ritchie	2	3	7	1	13	3.0
Laidlaw	43	12	14	0	69	3.2
Lang	22	25	15	0	62	3.8
Sinclair	37	57	5	1	100	4.7
All LC	118	109	55	2	284	4.1
Middle-class						
Nicoll	16	5	4	0	25	1.6
MacDougall	13	11	2	0	26	2.7
Muir	12	4	0	0	16	3.5
Gibson	15	21	2	0	38	4.5
MacGregor	35	10	0	0	45	5.8
Menzies	11	20	0	0	31	6.1
All MC	102	71	8	0	181	3.6

In types of subordination, the most obvious difference is in the use of subordinate noun clauses, for which the frequency is 12.0 in the middle-class group, compared with 5.8 in the lower-class group, as shown in Table 7–11. Since the use of subordinate noun clauses is more characteristic of middle-class speech, it is not surprising that Sinclair and MacDougall show the highest frequency in their respective groups.

There is also a range of *that*-deletion in subordinate noun clauses. Since *that* is an optional feature, it may occur as in 6a or be omitted as in (6b):

(6)   a. HG272   and naebody kent
         273   *that* I was daeing it
      b. WR734   I think
         735   ø they'd mair respect for their parents in them days
                ken

As can be seen from Table 7–12, *that*-deletion is more frequent among lower-class speakers, but there is overlap between the two groups. *That*-deletion is apparently not strongly conditioned by social class; nor does it appear to be affected by the interview situation, since both in individual interviews and in the interviews as a whole the frequency of *that*-deletion is more or less

Table 7.11. Subordinate Noun Clauses

Respondent	/1,000 wds.	n
Lower-class		
Rae	1.4	7
Lang	3.6	58
Ritchie	5.2	23
Gemmill	5.4	53
Laidlaw	5.9	78
Sinclair	8.6	183
All LC	5.8	402
Middle-class		
Nicoll	6.6	101
MacGregor	8.6	66
Menzies	12.2	62
Gibson	12.8	109
Muir	17.8	82
MacDougall	19.7	193
All MC	12.0	613

Table 7.12. *That*-Deletion

Respondent	Percentage
Middle-class	
MacGregor	47 (31/66)
MacDougall	64 (123/193)
Nicoll	70 (71/101)
Gibson	72 (79/109)
Menzies	87 (54/62)
Muir	87 (71/82)
All MC	70 (429/613)
Lower-class	
Gemmill	74 (39/53)
Sinclair	75 (138/183)
Ritchie	87 (20/23)
Laidlaw	90 (70/78)
Lang	97 (56/58)
Rae	100 (7/7)
All LC	82 (330/402)

constant. It is probably affected to some extent by topic and genre, but it was not possible to establish this quantitatively.

Following the example of Thompson (1988), however, certain contexts of *that*-deletion were examined. As can be seen from Table 7–13, first-person-singular subjects favor *that*-deletion. The middle-class figures are very similar to those Thompson found in her sample. (The lower-class speakers have a higher rate of deletion in all contexts.) She also found that *that*-deletion is most frequent after the verb *think*. The figures for *I think* are given in Table 7–14. Thompson also points out that the nature of the verb phrase can affect *that*-deletion. In the Ayr interviews, several factors inhibit *that*-deletion. The one that affects the greatest number of examples is tense. As can be seen from

Table 7.13. *That*-Deletion: *I* versus Other Subjects (%)

Class	I		Other	
	ø	*That*	ø	*That*
Middle	80 (271)	20 (67)	43 (65)	57 (86)
Lower	91 (159)	9 (16)	68 (96)	32 (46)
All speakers	84 (430)	16 (83)	55 (161)	45 (132)

Table 7.14. *That*-Deletion: *I Think*

Class	Percentage	
	ø	*That*
Middle	88 (143)	12(20)
Lower	97 (61)	3 (2)
All speakers	90 (204)	10 (22)

Table 7–15, *that*-deletion is much less likely after past tense verbs. Other factors inhibit *that*-deletion even more strongly; but the actual number of occurrences is small, and only figures for the whole group are given in Table 7–16. (Two other factors cited by Thompson showed no inhibiting effect on deletion—the presence of an indirect object [84% deletion] and the progressive form of the main verb [80% deletion]. There were too few examples of a following adverbial to be significant, but in those few there was also *that*-deletion.)

Of the other major subordinate clauses four are used more frequently by middle-class speakers, and three are found more frequently in the lower-class interviews. These two groups are shown in Tables 7–17 and 7–18. While the individual differences are small, as a group these clauses have a frequency of 9.1 per thousand words in the middle-class interviews but only 6.1 in the lower-class interviews. Table 7–18 shows the types of subordinate clause that are more frequent in the lower-class interviews. In this case clauses of these types occur with a frequency of 9.0 per thousand words in the lower-class interviews but only 7.0 in the middle-class interviews.

The differences are large enough to raise the question whether there is anything that the clauses in each group have in common that might account for this social class difference. The first group (sb, sc, si, and sq) are used for the purposes of explanation, qualification, and speculation; while the second group are used more for description. As might be expected, in MacDougall's interview there is a higher frequency of the first group (15.3) and a lower

Table 7.15. *That*-Deletion: Effect of Tense (%)

Class	Present		Past	
	ø	*That*	ø	*That*
Middle	82 (261)	18 (57)	53 (61)	47 (54)
Lower	91 (142)	9 (14)	74 (68)	26 (24)
All speakers	85 (403)	15 (71)	62 (129)	38 (78)

Table 7.16. *That*-Deletion: Other Factors Inhibiting Deletion (%)

Type of Verb	ø	*That*
Finite	75 (582)	25 (197)
Infinitive	33 (9)	67 (18)
Active	74 (590)	26 (209)
Passive	14 (1)	86 (6)
Affirmative	74 (574)	26 (199)
Negative	52 (17)	48 (16)

frequency of the second group (9.4). Sinclair's interview shows exactly the reverse, with a frequency of 6.3 in the first group and 12.9 in the second. Since MacDougall's interview has very little description or narrative while Sinclair's is rich in both, it seems likely that the cause of the difference in the group scores is primarily a matter of genre and topic. Some limited support for this view comes from the interview with Nicoll. Nicoll's interview has the most narratives of any of the middle-class interviews; and he has approximately the same frequency of each group of subordinate clauses, 6.0 and 6.3, respectively. Thus, the difference between the two social class groups in this respect probably follows from the fact that there are more narratives and descriptions in the lower-class interviews.

There is one further difference in clauses of comparison, which in the lower-class interviews sometimes have *what* after the conjunction, as shown in (7):

(7) a. AS1539   there'd be far more production
        1540   *than what* they're getting with the system
     b. AS3168   she's mair with the weans I daresay
        3169   *than what* the man is at times
     c. WL2437   of coorse the traffic wasnae as strong
        2438   *as what* it is noo

Table 7.17. Subordinate Clauses Used More Frequently by Middle-Class Speakers (/1000 wds.)

Class	sb	sc	si	sq
Middle	3.2 (164)	.6 (31)	3.2 (164)	2.1 (105)
Lower	1.9 (135)	.3 (23)	2.3 (158)	1.5 (107)

Note: sb = subordinate clause of reason, sc = subordinate clause of concession, si = conditional clause, sq = embedded question.

Table 7.18. Subordinate Clauses Used More Frequently by Lower-Class Speakers (/1000 wds.)

Class	sa	sp	st
Middle	1.7 (89)	.4 (20)	4.8 (246)
Lower	2.1 (144)	.8 (57)	6.1 (427)

sa = subordinate clause of comparison introduced by *as*, sp = subordinate adverbial clause of place, st = subordinate adverbial clause of time.

Sometimes *nor* occurs in place of *than*, as in the examples in (8):

(8) a. MR715   well it was better then
      716   *nor what* I think
      717   it is noo
  b. HG316   you couldnae get any mair *nor* two pound

These usages are found only in the lower-class interviews; and this use of *nor* occurs only in the interviews with the two oldest speakers, Gemmill and Ritchie.

There are some differences in the use of infinitives (in), gerunds (ge), and present participles (pp). Since their distribution differs from finite clauses, the measure for comparison will not be percentages but the frequency of these structures per thousand words. The frequencies for both social class groups are given in Table 7–19. It can be seen from Table 7–19 that the greatest differences are in infinitives, with the middle-class speakers being much more likely to use an infinitive than the lower-class speakers, with the exception of Sinclair. There is a much smaller difference in the use of gerunds and essentially none in the use of present participles. Thompson has argued that what she calls "the detached participle" is "characteristic of literary rather than conversational English" (1983, 46). In one literary text she has found a

Table 7.19. Frequency of Nonfinite Structures

Class/Respondent	IN		GE		PP	
	/1,000 wds.	n	/1,000 wds.	n	/1,000 wds.	n
Middle-class	13.1	629	5.5	265	2.8	131
Lower-class	9.8	625	4.8	306	2.6	166
Sinclair	13.4	275	6.1	125	2.0	40
LC w/o Sinclair	8.1	354	4.2	181	2.9	126

frequency of 7.4 participial phrases per thousand words. The frequency in the Ayr interviews is 2.5 per thousand words, which would appear to confirm Thompson's view that such phrases are less characteristic of conversation. Their low frequency may be a factor in the lack of variation in their use. Infinitives, on the other hand, occur with an overall frequency of 10.5 per thousand words. The higher frequency among the middle-class speakers is quite marked, with MacDougall showing a frequency of 17.9 per thousand words. The overall frequency of gerunds is much lower, and the differences between the two social groups is much less. MacDougall has a frequency of 7.2 gerunds per thousand words, which is exceeded only by Gibson, whose frequency is 7.5. As in other cases, Sinclair's figures of 13.3 for infinitives and 5.9 for gerunds are more consistent with the middle-class pattern than with the figures for the other lower-class speakers.

Infinitives are the surface forms of a number of very different syntactic relationships. For the purposes of the following analysis, three kinds of infinitives must be distinguished: (i) infinitives in construction with another verb, an adjective, or a noun, as shown in (9); (ii) infinitives of purpose, as illustrated in (10); and (iii) free infinitives, as illustrated in (11).

(9)   a. WG1063   I'm trying
            1064   *to think* of the older people
      b. DN1805   my father loved
            1806   *to tell* the story
      c. AS2502   that they would allow one of our men
            2503   *to go off* half an hour earlier
      d. EL1116   because we weren't allowed
            1117   *to have* it
      e. DN1555   and I would persuade two or three fellows
            1556   *to use* their cars
      f. WL800   I always seemed
            801   *to be* the ambulance man sort of thing
      g. IM412   it was difficult
            413   *to play* rugby initially
      h. WL340   you see it was an offence
            341   *to go doon* with cigarettes in your pocket

Although the examples in (9) show different syntactic relationships, the infinitives have a common syntactical dependence on a constituent in a higher sentence. This is different from the examples in (10) and (11).

(10)   a. EL1190   so they sent me
             1191   *to help* them

  b. DN1931   she was lying
     1932   *to defend* her boy
  c. WL816   so I went oot on to the roadheid
     817   *to see aboot* this wid
  d. JM1249   and he sent the janitor out on his bike
     1250   *to see*

The examples in (10) are all examples of purpose infinitives, equivalent to subordinate clauses of purpose. There is no single constituent in the higher sentences that is in construction with the infinitive.

(11) a. NM465   I did take Latin
         466   *to begin* with
      b. DN2208   however *to get back* to France
      c. AS1778   oh *to give* you the best example
      d. AS1855   *to tell* you the truth

The examples in (11) are not infinitives of purpose, but neither are they in construction with any single constituent in the main clause.

There are some social class differences in the use of these three kinds of infinitives, as can be seen in Table 7–20. As can be seen from Table 7–20 the proportion of infinitives of purpose and free infinitives is twice as high in the lower-class interviews. Within the category of infinitives in construction with verbs there are few differences in the kinds of constructions, though there is a curious difference in the case of infinitives as in 9a and b, where the underlying subject of the infinitive is identical with the subject of the higher verb. In the middle-class interviews 43% of the examples are of first-person-singular *I*—in the lower-class interviews only 17%.

One other feature of infinitives shows social stratification. There are 21 examples of *for to* infinitives, all in the lower-class interviews. More than half

Table 7.20. Distribution of Types of Infinitives

Type of Infinitive	Lower-Class		Middle-Class	
	%	n	%	n
i/c with verbs	46	286	62	387
i/c with adjectives	8	48	13	82
i/c with nouns	11	69	10	60
Purpose	29	183	12	73
Free	7	43	4	23

(12) of these can be interpreted as infinitives of purpose, as in the examples in (12):

(12)   a. EL478   he made me a schoolcase—a case
          479   *for to carry* my schoolbooks in
       b. HG1207 you don't need to faw ten thousand feet
          1208   *for to get killt*
       c. AS544   and aw these hutches was sent out to the bing
          545   *for to be couped* you see

The remaining 9 examples are much harder to interpret as infinitives of purpose, as can be seen in the examples in (13):

(13)   a. HG207   you werenae allowed at this time
          208   *for to go* and take another job on
       b. MR539   but my own brothers was all too old
          540   *for to go*
       c. MR294   we had the clear road
          295   *for to play* on

The use of *for to* infinitives may be linked to age. The two youngest lower-class speakers, Rae and Lang, have no *for to* infinitives in their interviews; and the two oldest, Gemmill and Ritchie, use them most frequently. In the interview with Gemmill 13% of the infinitives are *for to* infinitives, and Ritchie has 10% *for to* infinitives. Laidlaw and Sinclair have 2.6% and 1.7%, respectively. The use of *for to* infinitives thus correlates exactly with age and suggests that this construction may be about to disappear from Ayr speech.

To sum up, middle-class speakers exhibit a greater use of subordination in general and a more frequent use of subordinate noun clauses in particular. Middle-class speakers also tend to use infinitives more frequently and, to a much less marked extent, gerunds. There are differences in the use of other subordinate clauses, with the middle-class speakers using more subordinate clauses of reason, concession, and condition and embedded questions and the lower-class speakers using more subordinate clauses of comparison, place, and time; but these differences are probably a reflection of differences in topic and gender between the two sets of interviews. No clear social class differences were found in the use of passives or present participles, in clause length, or in the frequency of *that*-deletion. A range of variation was found within each social class group and one lower-class speaker, Sinclair, was found to display certain characteristics of middle-class speech in his interview.

8

Lexical Features

Of all aspects of language, the lexicon is the most difficult to characterize from a small corpus. The choice of words is constrained by topic. Thus, it is not surprising that the interview with MacDougall contains technical terms from academic subjects that do not occur in any of the other interviews; nor is it unexpected that Sinclair's interview contains various terms referring to coal mining—as does Lang's to some extent. (Rae, on the other hand, says very little about coal mining, perhaps because he was the only one of the three who was still a miner at the time of the interview and may have had a less nostalgic view of life in the pit.) Even at its most extensive, however, the evidence on any section of the lexicon is fragmentary and not in the same category as, for example, Macafee's investigation of lexical items in Glasgow (Macafee 1988), Wright's work on the language of coal mining (Wright 1972, 1974), or the responses to the Linguistic Survey of Scotland's postal questionnaire (Mather and Speitel 1975–86, vols. 1, 2). Even the morphological evidence contains many gaps; but it is important because morphological differences, particularly irregular (or "strong") verb forms, are often very salient markers of social class differences.

Verb Morphology

In the lower-class interviews, as Macafee (1980) and James Milroy (1981) have pointed out in other situations, there is a tendency to reduce irregular verb paradigms to a more regular pattern by using the same form for the preterite (past tense) and the past participle. The most frequent type is where the past participle is used for the preterite, for example, *come, done, run, seen,* and *ta'en.* The frequencies of these forms in the lower-class interviews

107

Table 8.1. Past Participles As Preterites (%)

Respondent	Come	Done	Run	Seen	Ta'en
Gemmill	57 (17/30)	0 (0/16)	0 (0/0)	100 (6/6)	0 (0/14)
Laidlaw	17 (5/29)	0 (0/13)	50 (1/2)	15 (2/13)	13 (1/8)
Lang	54 (22/41)	57 (17/30)	50 (2/4)	100 (7/7)	53 (10/19)
Rae	50 (4/8)	50 (3/6)	0 (0/0)	100 (1/1)	0 (0/3)
Ritchie	50 (5/10)	25 (1/4)	100 (1/1)	100 (3/3)	0 (0/0)
Sinclair	67 (31/46)	35 (7/20)	100 (3/3)	10 (1/10)	22 (2/9)
All LC	50 (82/164)	31 (28/89)	70 (7/10)	50 (20/40)	25 (13/53)

are shown in Table 8–1. Table 8–1 shows that for some speakers *seen* is categorical for the preterite, though the numbers are small. Table 8–2 shows that for some speakers *went* is categorical as the past participle of *go*. (Lang has also one example of *went* as an infinitive: WL689/90 *see if he'd tried to went oot*.) Sinclair is the only one who shows signs of what Milroy calls a "flip-flopped" paradigm (1981, 14):

(1)   AS1613   I *seen* them from management and everything doon
          250   I've *saw* hundreds
          251   that was there

However, this is Sinclair's sole use of *seen* as a preterite, and the use of *saw* as a past participle is the more frequent (5/6, or 83%). Since Sinclair is the only speaker who uses *saw* as a past participle and since his interview shows many examples of accomodation to the interview situation, it is possible that this is a form of hypercorrection. Tentative support for this view is the fact that the only other speaker to use past tense forms (other than *went*) for the past

Table 8.2. *Went* As a Past Participle

Respondent	Percentage
Ritchie	0 (0/0)
Laidlaw	29 (2/7)
Rae	67 (2/3)
Gemmill	100 (2/2)
Sinclair	100 (3/3)
Lang	100 (4/4)
All LC	68 (13/19)

participle is Ritchie, who also shows signs of accomodation; but there are only two examples, *took* and *did*.

Done never occurs as the past tense of auxiliary *do* in negatives, questions, or for emphasis. It is used almost equally as a main verb (54%) and as an anaphoric pro-verb (46%). Examples are given in (2):

(2)   a. WL1330   I never *done* many lessons
        b. AS515   but they just *done* a wee job
        c. WL488   it was the wrang thing
             489   to dae mind you
             490   but still I *done* it
             491   I *done* that

The other way to regularize the paradigm is to treat the verb as one that forms its past tense and past participle by the addition of /t/ to the base form. The most frequent example is *tell*, as seen in Table 8–3. The salience of this item probably accounts for its absence in the interviews with Sinclair and Ritchie and also for the fact that the past tense uses of *told* in Laidlaw's interview occur in the first quarter of the interview. The other examples of this type of regularized paradigm are past tense *selt* (one example in Gemmill's interview) and *gied* (one example in Gemmill's interview and two in Laidlaw's). Laidlaw also uses *gave* four times; Gemmill has no examples of *sold* or *gave*.

Other isolated examples are *fotched* (past tense of *fetch*), *driv* (past tense of *drive*), *brung* (past participle of *bring*) and *gotten* (past participle of *get*), *writ* (past tense of *write*), *blew* (past participle of *blow*), *stole* (past participle of *steal*), and *bate* (past tense of *beat*).

Table 8.3. Proportion of Telt As Past Tense and Past Participle

Respondent	Telt (Past Tense)	Telt (Past Participle)
Gemmill	100 (1/1)	100 (1/1)
Laidlaw	60 (3/5)	67 (4/6)
Lang	100 (7/7)	100 (1/1)
Rae	100 (3/3)	0 (0/0)
Ritchie	0 (0/1)	0 (0/0)
Sinclair	0 (0/6)	0 (0/5)
All LC	61 (14/23)	46 (6/13)

Number Marking on Nouns

It has been claimed (Trudgill, Edwards, and Weltens 1983, 32) that in British dialects "it is an almost universal rule that, after numerals, nouns of measurement and quantity retain their singular form." This is not the case in the Ayr interviews. In the lower-class interviews there are 90 examples, and 41 (46%) have an inflected plural. *Minute, day, week, shilling, inch,* and *yard* are always inflected. The percentage of inflected plurals for the other nouns of measurement and quantity are as follows: *pound* (both money and weight) 89%, *month* 86%, *year* 68%, *ton* 50%, and *mile* 17%.

Prepositions

An important difference between the lower-class and middle-class interviews that cannot be easily quantified is the use of prepositions. The most obvious is the use of *fae* for *from:*

(3)   a. MR030   she came *fae* Patna direction
      b. WL1607   we got eggs *fae* him
      c. AS509   who were retired *fae* underground
      d. MR011   I'm seventy-year *fae* November

Fae is also used as a conjunction:

(4)   a. MR804   by it's a while
         805   *fae* I heard that
      b. AS1712   the first time I ever was idle
        1713   *fae* I left the school

The next-most-noticeable preposition is *with,* which is used in a variety of meanings:

(5)   a. HG1394   oh I never would go back *with* a train again
      b. WL1948   oh aye treveled through the fields *with* [i.e., on] his bike
      c. WL1750   and he went doon *with* an auld car you ken
      d. MR471   when the lady
         472   that I worked *with* [i.e., for]
         473   died suddent
      e. WL302   well there a difference *with* [i.e., between] the stamp works and the pits

Other prepositions with a variety of uses are *to* and *on:*

(6)   a. MR148   he worked *to* Wilson of Troon
      b. WR153   I'm laboring *to* a bricklayer
      c. WL1599   and we went *to* the milk in the morning [i.e., deliv-
                   ering the milk]
      d. AS2155   never was *to* Ireland to this day
      e. WL1518   I'll no go *to* [i.e., for] the cheese
(7)   a. HG1166   I never thocht *on* anything
      b. AS3685   and maybe youse are talking *on* different things
      c. EL1277   he was married *on* one of the Keilers of Dundee
      d. EL3598   I'm awful glad
         3599   I'm no waiting *on* you
      e. AS2694   I never ever thought
         2695   *on* going abroad at all
      f. WL2970   we go
         2971   *on* ferreting on a Sunday

There is also some alternation between *out of* and *out:*

(8)   a. HG224   which got me *oot of* the army the second war
         225   which got me *oot* the army again in a way
      b. MR076   I never was *oot of* that hoose
         463   I was glad
         474   to get *oot* it

Lexical Variety

The availability of a commercially developed computerized concordance program (WordCruncher) made possible word counts of a kind that would have
been extremely time-consuming (see below) to do otherwise. The problem
with such automatic programs, however, is that they do not distinguish similar
word forms that can be different parts of speech; that is, for example, the
program does not differentiate between *like* as a verb or a preposition, or *well*
as an adverb from *well* as a discourse marker. Although the number of ambiguous items is a small proportion of the total, the relative opacity of English
word classes makes an exhaustive listing of parts of speech from decontex-
tualized word forms a hazardous business. However, there are unambiguous
items such as the pronouns and most adverbs; and the value of overall frequencies in drawing attention to interesting patterns of distribution has been
shown in Chapter 6. There are also more global lexical characteristics of the
interviews that can be calculated.

Table 8.4. Type–Token Ratios: Complete Interviews

Respondent	Number of Types	Number of Tokens	TTR
Lower-class			
Ritchie	800	4,373	.183
Rae	1,033	5,030	.199
Gemmill	1,505	9,761	.154
Laidlaw	2,026	13,189	.154
Lang	1,840	16,255	.113
Sinclair	2,027	21,163	.096
All LC	2,983	69,771	.043
Middle-class			
Muir	811	4,573	.177
Menzies	1,090	5,110	.213
MacGregor	1,320	7,673	.172
Gibson	1,603	8,510	.188
MacDougall	1,551	9,764	.159
Nicoll	1,986	15,268	.130
All MC	3,141	50,898	.062

The *type–token ratios* (TTRs)[1] for the complete interviews can be seen in Table 8–4. It is important to be clear what the figures in Table 8–4 represent. The types and tokens are not "words" in any linguistic sense but strings of characters preceded and followed by a space. Thus, contracted forms such as *I'm, could've,* and *there's* are counted as single items. Moreover, this is not a lemmatized concordance; for example, *break, breaks, broke* and *broken* are distinct "types," as are all the different forms of the verb *be.* Moreover, there is no way to distinguish forms that are homographic (e.g., *can* as a noun or an auxiliary or *mean* as a verb or an adjective). However, since the same coding principle applies equally to all the interviews, figures are presumably adequate for comparative purposes.

Unfortunately, TTRs for samples of unequal length are not comparable (Miller 1951, 123), since the ratio gets smaller as the size of the sample gets larger. For this reason, in order to make a meaningful comparison between the two sets of interviews, I analyzed samples limited to the length of the shortest interview (Ritchie's). The first approximately 4,370 words from each inter-

---

[1]The type-token ratio (TTR) is calculated by treating each distinct word form as a type and each occurrence of that word form as a token. The TTR is the number of types divided by the number of tokens. The higher the ratio, the more varied the vocabulary. It is necessary to specify the TTR in terms of word forms because that is what the program counts.

view were extracted and analyzed. The results are shown in Table 8–5. It can be seen from Table 8–5 that while the vocabulary in the middle-class interviews is slightly more varied than that in the lower-class interviews, the differences are not great and all the ratios (with two exceptions—Sinclair and Gibson) fall within a single range. The individual comparisons are also interesting. MacDougall and Rae, with very different educational backgrounds, have identical TTRs, as have Laidlaw and Menzies. As in so many other features, Table 8–5 shows that generalizations about social class differences in vocabulary should be treated with caution.

As was shown in Table 8–4, in the middle-class interviews there are 3,141 word types, compared with 2,983 in the lower-class interviews. However, these two sets are intersecting; 1,771 word types are common to both sets of interviews. The question then arises whether there are characteristics of the word types that are unique to one or the other set of interviews. One aspect that shows a significant difference is the number of polysyllabic words. For this count, a polysyllabic word was defined as one having three or more syllables; no attempt was made to distinguish between words of Romance and Germanic origin. As can be seen from Table 8–6, a much higher proportion of polysyllabic words is found in the word types unique to the middle-class interviews. It is important to note, however, that it is the shared vocabulary

Table 8.5. Type–Token Ratios: Equal Samples

Respondent	Number of Types	Number of Tokens	TTR
Lower-class			
Sinclair	746	4,376	.170
Lang	784	4,376	.179
Ritchie	800	4,373	.183
Gemmill	875	4,364	.201
Rae	942	4,371	.215
Laidlaw	978	4,377	.223
All LC	2,753	26,190	.105
Middle-class			
Muir	783	4,370	.179
MacGregor	861	4,360	.197
Nicoll	902	4,379	.206
MacDougall	941	4,372	.215
Menzies	979	4,376	.224
Gibson	1,053	4,370	.241
All MC	3,008	26,180	.115

Table 8.6. Word Types

Class Character	All n	Polysyllabic n	%
Unique to middle-class	1,370	565	41
Unique to lower-class	1,213	330	27
Common to both	1,771	240	14
Total	4,354	1,135	26

that has the lowest proportion of polysyllabic words and that more than a quarter of the words found only in the lower-class interviews are polysyllabic. In Table 8–7 it may be obvious to some that the words listed are characteristic of one social class group rather than the other, but it is hard to see objective criteria for distinguishing between the two lists. There are many other examples and, given the topics actually discussed in the individual interviews, it is sometimes possible to see why a particular term should have arisen in one interview rather than another; but the reason is generally not to do with social class in any simple correlation. It is true that nominalizations are more frequent in the middle-class interviews, as Table 8–8 shows. The difference, however, is not great.

In order to investigate other lexical differences between the two social class groups, two interviews were chosen for detailed analysis. The interviews with

Table 8.7. Words Found Uniquely in Middle-Class or Lower-Class Interviews

Lower-Class	Middle-Class
atrocious	alternate
combustible	correlation
degradation	desolation
expulsion	equivalent
fanatical	fabulous
harmony	hypocrisy
illiteracy	inertia
materialize	mobility
notorious	naturalistic
parallel	penetrate
spasmodic	spontaneity
transition	technicality
vulgarity	vernacular

Table 8.8. Frequency of Nominalizations

Class	Types	Tokens	Tokens/1,000 wds.
Lower	47	80	1.15
Middle	67	111	1.57

Gemmill (9,761 words) and Gibson (8,510 words) are similar both in length and in other characteristics (e.g., average length of response to questions, proportion of narrative) and long enough to provide a suitable sample for comparison. (MacDougall's interview is almost identical to Gemmill's in length but very different in nature, so it provides a less appropriate comparison.) All the content words (with the exception of proper nouns) in both interviews were indexed in the categories noun, verb, adjective, and adverb; and the word-processing program sorted and listed them alphabetically by page number. The items were then counted by hand to provide a total of types (lexemes) and tokens (word forms). It was a time-consuming and frustrating process. As anyone who has attempted a similar task will know, the coding process itself is fraught with problems. Major problems involve compound words, including phrasal verbs, idioms, and formulaic phrases. There were also problems with identifying word class, particularly with adverbs and prenominal modifiers. Arbitrary decisions were made at many points, but every attempt was made to maintain consistency. The figures are therefore mainly intended for the comparison between the two speakers, as shown in Table 8–9.

Perhaps the most striking feature of Table 8–9 is the similarity between the two sets of figures. Given the considerable differences in social background, age, and experience and also the differences in the topics discussed, the overall proportion of form classes and their variety are remarkably similar.

Table 8.9. Type–Token Ratios for Gemmill and Gibson

Part of Speech	Gemmill			Gibson		
	Types	Tokens	TTR	Types	Tokens	TTR
Nouns	425 (49%)	837 (39%)	.51	474 (45%)	877 (38%)	.54
Verbs	261 (30%)	828 (39%)	.31	278 (26%)	788 (34%)	.35
Adjectives	104 (12%)	208 (10%)	.50	218 (20%)	388 (17%)	.56
Adverbs	71 (8%)	269 (12%)	.26	95 (9%)	273 (12%)	.35
All	861 (100%)	2,152 (100%)	.40	1,065 (100%)	2,326 (100%)	.46

Note: The total TTRs are very different from those given in Table 8–4 because they are for content words only.

Table 8.10. Proportion of Lexical Items According to
Syllable Structure

Part of Speech/Respondent	Monosyllabic	Disyllabic	Polysyllabic
Nouns	%	%	%
Gemmill	52	34	14
Gibson	38	37	25
Verbs			
Gemmill	69	28	3
Gibson	55	35	10
Adjectives			
Gemmill	47	33	20
Gibson	33	33	33
Adverbs			
Gemmill	48	35	16
Gibson	23	41	36
All categories			
Gemmill	55	33	12
Gibson	40	36	24

Overall and in every category, Gibson has a more varied vocabulary, as shown by the higher TTRs; but once again, the differences are not great.

The frequencies of individual items show that there are considerable differences between the two interviews. As a general characterization, it would be reasonable to say that the most frequent lexical items in Gemmill's interview tend to be more toward the "concrete" end of the continuum and Gibson's more toward the "abstract" end, although these notions are extremely hard to define precisely. Since there is a relationship between number of syllables and abstractness in English because of the tendency to use items from the Romance part of the lexicon for abstract qualities, a count of the lexical items was made in three categories: monosyllabic, disyllabic, and polysyllabic. The comparative figures for the two interviews are given in Table 8–10. As Table 8–10 shows, the proportion of polysyllabic lexical items is twice as high in Gibson's interview, while the proportion of monosyllabic lexical items is approximately a third higher in Gemmill's interview. Given the difference in education and social class background—not to mention topics discussed in the interviews—this is what one would expect. It is interesting, however, that despite the considerable differences shown in Table 8–10, the overall configurations shown in Table 8–9 should be so similar. It is further evidence that in addition to the kind of salient social class differences shown in pronunciation, verb morphology, and negation, there are also more subtle differences that are linked more to the use of language than to the underlying system.

I have shown that in addition to the more systematic aspects of language variation in phonology and syntax, there are patterns of social class variation in the lexicon also. Some of these differences are highly salient though not necessarily very frequent (e.g., verb morphology). Others, such as the kind of lexical items used, show global differences when the two sets of interviews are considered as a whole; but finer analysis reveals that such differences are probably more closely linked to topic than to social class.

9

The Expression of Intensity

Recently, Hymes has claimed, "The abilities of individuals and the composite abilities of communities cannot be understood except by making 'verbal repertoire,' not 'language,' the central scientific notion" (1984, 44). Hymes goes on to suggest that this will require "a mode of description in linguistics which can address the organization of linguistic features in styles, that is, in ways which cut across the standard levels of linguistic structure" (ibid.). In this chapter I will deal with just one feature of the interviews, that which Labov refers to as *intensity:* "At the heart of social and emotional expression is the linguistic feature of intensity. It is a difficult feature to describe precisely. Intensity by its very nature is not precise: first, because it is a gradient feature, and second, because it is most often dependent on other linguistic structures" (1984, 43). Labov deals primarily with universal quantifiers in his paper, although there are some very important references to other forms of signaling intensity. I shall be looking at the wider range of expressions of intensity, and I shall use as a framework the differences between the two social groups in their expression of intensity, though I shall argue later that the differences between the groups may be largely due to differences in topic and genre. I shall deal with three ways in which intensity is expressed in the interviews: (1) word order, (2) adverbs, and (3) prosodic features.

Word Order

The unmarked order for a declarative clause is subject–verb–complement. Of the 10,568 declarative clauses in the total sample, 10,038 (95%) have this order. Any deviation from this order consequently has the rhetorical force of bringing some constituent into greater prominence. Middle-class and lower-class speakers use variation in word order to approximately the same extent (middle-class 4.3%, lower-class 5.5%), but there is a difference in the proportion of each structure used. The most common is adverbial fronting, as illustrated in (1):

(1) a. AS1918   you see *for that pool* you could buy your pint of
                  beer your stout or a strong ale or that
     b. WL1892   *away* they went to their bags you ken
     c. JM858   *for the rest of the staff* it was very difficult
     d. AM656   *just one evening a week* I went to a night class

Adverbial fronting accounts for 84% of the marked word orders in the middle-class group and 51% in the lower-class interviews. In other words, the middle-class speakers seldom deviate from the unmarked word order in any other way than by adverbial fronting. (As was pointed out in Chapter 7, there is limited use of the passive construction in either group.)

The most striking difference between the two groups, however, is that the lower-class speakers make more frequent use of such highlighting devices as clefting, left and right dislocation, and NP-fronting, as was shown in Chapter 6. (For convenience, some of the material presented in Chapter 6 will be repeated here.) The most common of these devices is left dislocation, in which an initial noun phrase is referred to later in the same clause by an anaphoric expression, usually a pronoun. There are 63 clear examples in the interviews, 56 (89%) in the lower-class interviews and 7 (11%) in the middle-class interviews. The most frequent cases are those in which the focused NP is the subject of the clause:

(2) a. EL397   but *my own family they*'ve had a lot of leeway
     b. HG621   *my auld lady she* was tying tae
     c. WL1237   *Mr Patterson he* was a gentleman
     d. AS1951   and *the rooms* oh *they* were very small

Sometimes the focused NP can be quite complex:

(3) a. EL3696   *that young brother of mine*
         3697   I'll tell you
         3698   *he* buried a carpet beater in the garden
     b. HG566   and *the fermer*
         567   *that I was wi at that time*
         568   *he* was bate oot o the ferm

There are no clear examples of left dislocation of subjects in the middle-class interviews. The few possible examples show a broken intonation pattern that is very different from the lower-class examples:

(4)   a. WG194   and after the war *my brother*
            195   who was three years younger
            196   than—I am
            197   ah—*he* was interested in photography
      b. JM871   but—eh—*the woodwork teacher and the domestic teacher*
            872   when they were working
            873   *they* had a terrible time

The focused NP can also be a direct object or belong to a prepositional phrase:

(5)   a. HG1386   oh *photos* in there I've hunners of *them* of London
      b. AS1324   *correct temperature* you get *it* through pressure et cetera and that
      c. WG1279   *the lower characters* make *them* Scots
      d. NM1309   *Postman's Knock and similar little things*
            1310   don't tell me
            1311   the boys hated *these*

The clear examples of left dislocation in the middle-class interviews are of this kind.

Right dislocation is slightly less frequent in the lower-class interviews, and there is only one rather dubious example in the middle-class interviews. Right dislocation is the mirror image of left dislocation, with an NP occurring in clause-final position and coreferential with a pronoun earlier in the same clause. The coreferential pronoun can be either a subject or an object:

(6)   a. WL1373   in fact *he* offered me a job *Mr Cunningham*
      b. WR786   *she* was a very quiet woman *my mother*
      c. EL3786   and we'll play a trick on *her old Mrs McNaught*
      d. HG772   I was asking John
      d. HG773   if he ever heard of *it Cabbies Kirk*

As pointed out in Chapter 6, when the focused NP is itself a pronoun, the coreferential pronoun must be in subject position:

(7)   a. WL1315   *he* was some man *him*
      b. HG859   but *she* was a harer *her*
      c. WL1413   oh and *it* was a great thing *this*
      d. WL2159   of coorse *it*'s a big haw *that*

The examples in Item 7 show right dislocation as a means of adding emphasis where—at least in examples 7a and b—middle-class speakers would have been more likely to use emphatic stress. A similar manner of focusing on pronouns occurs with clefting.

Clefting is a construction with *it* as a dummy subject and a form the verb *be* followed by an NP and a relative clause. There are 48 examples in the total sample, 37 (77%) in the lower-class interviews and 11 (23%) in the middle-class interviews.

(8) a. JM701   actually it was Dunfermline
        702   that I went to for my interview
     b. AM1680   it's only Simpson
        1681   that does anything at all in that line
     c. HG336   by that time it was half a croon
        337   you were paying
     d. EL1836   it's a queer man and wife
        1837   that doesnae have an argument

In the lower-class interviews, clefting is an environment that allows a ø relative marker even in subject position:

(9) a. WL847   it was Jimmy Brown
        848   ø was the fireman
     b. WL1862   and it was my mother
        1863   ø was daeing it

A ø relative marker, however, is never possible when the focused NP is a pronoun:

(10) a. WL1411   it was her
        1412   that got them
     b. EL660   no it's me
        661   that's het
     c. EL3118   it was him
        3119   that led the band
     d. MR639   it's them
        640   that's running it now

This focusing on pronouns in right dislocation and clefting also occurs in the construction that was called *demonstrative focusing* in Chapter 6. This consists of a demonstrative pronoun followed by a form of the verb *be* and the object form of a pronoun with a complement that can be either an adjective, a past participle, a present participle, or an adverbial clause:

(11) a. WL1474   that was *him* idle
     b. EL394   and that was *you* shut in the house for a week
     c. WL2116   but that's *me* seen it
     d. WL1147   that's *us*
        1148   going for another game

    e. HG583   and that was *you*
       584   maybe slaving until eicht or nine at nicht
    f. WL1453   that was *you* till aboot six o'clock

Another highlighting device that often has the effect of leaving a pronoun in an exposed position is NP-fronting (similar to Y-movement). NP-fronting consists of moving an NP other than the subject to a position before the subject:

(12)   a. WR633   *an auld auld man* he was ø you ken
      b. AS2236   and *one of them* he had been out with ø once or twice
      c. JM077   *thirty shillings a quarter* we paid ø
      d. AM057   *primary school* I found ø a bit of a drag

Only in the lower-class interviews is a use of Y-movement found that has the effect of leaving a pronoun in clause-final position:

(13)   a. WL1875   *Davie Drummond* you caw'd him ø
      b. AS2612   *the Army and Navy Club* they called it ø

Although the pronoun is not stressed in final position, its relatively isolated position gives it slightly more prominence than it would have in the unmarked word order. For convenience, Table 6–4 is repeated below as Table 9–1.

As was pointed out in Chapter 6, in right dislocation, clefting, demonstrative focusing, and NP-fronting, the lower-class speakers bring personal pronouns into a kind of prominent position that is not found in the middle-class interviews. In addition, in the lower-class interviews, the object form of a personal pronoun occasionally occurs as the antecedent of a relative pronoun:

(14)   a. AS2990   *him*
       2991   that's now head of the union thing
      b. MR089   *them*
       090   that had wee families
      c. AS2323   and this was *them*
       2324   Britain was getting
      d. AS2990   *him*
       2991   that's now head of the union thing

Such personal pronoun antecedents are never found in the middle-class interviews where the alternative would be an indefinite pronoun:

(15)   JM819   *the ones*
               who had been working

Table 9.1. Highlighting Movement Rules

Type of Rule	Middle-class		Lower-class	
	n	/1,000 wds.	n	/1,000 wds.
Left dislocation	7	.14	56	.80
Right dislocation	1	.02	37	.53
NP-fronting	10	.20	30	.43
it-clefts	11	.22	57	.82
Demonstrative focusing	0	.00	23	.33
Total	29	.58	203	2.91

It is clear from the above examples that the lower-class speakers make use of a greater variety of word orders for focusing on parts of an utterance. A very different kind of situation is found in the use of adverbs.

Adverbs

Labov observes, "Intensity operates on a scale centered about the zero, or unmarked expression, with both positive (aggravated or intensified) and negative (mitigated or minimized) poles" (1984, 44). Intensity is conveyed through, among other ways, the use of adverbs in *–ly*. The middle-class speakers make much more frequent use of this device than do the lower-class speakers, both in the number and variety of such adverbs. There are 133 different adverbs in *–ly* in the total set of interviews; 20 are found only in the lower-class interviews, 72 occur only in the middle-class interviews, and 41 are common to both sets of interviews. In other words, of the adverbs in *–ly* that are found only in one set of interviews, 78% are found in the middle-class interviews. The middle-class speakers not only use more kinds of adverbs in *–ly*, they also use them more frequently—8.9 per thousand words compared with 2.6 per thousand words in the lower-class interviews (not counting *really*, which, as will be shown, is a special case). The differences are greater than this if Sinclair's interview is treated as a special case. As in so many other cases, Sinclair is closer to the middle-class group than to the other lower-class speakers. Sinclair uses adverbs in *–ly* with a frequency of 3.9 per thousand words; the frequency for the other lower-class speakers is 2.0 per thousand words. In other words, the middle-class speakers use adverbs in *–ly* more than three times as frequently as the lower-class speakers—four times as frequently if Sinclair's interview is excluded. The effect of these differences can be seen more clearly in Table 9–2. This is not a complete list of the adverbs in *–ly* in

Table 9.2. Frequency of Adverbs in –ly

Adverb	Middle-class		Lower-class	
	n	/1,000 wds.	n	/1,000 wds.
Absolutely	3	.06	0	.00
Actually	62	1.22	17	.24
Basically	9	.18	1	.01
Certainly	14	.28	10	.14
Completely	7	.14	2	.03
Consciously	2	.04	0	.00
Deliberately	5	.10	1	.01
Distinctly	2	.04	0	.00
Especially	12	.24	4	.06
Eventually	10	.20	7	.10
Exactly	8	.16	1	.01
Extremely	5	.10	0	.00
Fairly	9	.18	0	.00
Finally	6	.12	0	.00
Generally	9	.18	3	.04
Highly	3	.06	0	.00
Immediately	4	.08	0	.00
Initially	3	.06	0	.00
Literally	5	.10	0	.00
Mainly	2	.04	0	.00
Necessarily	6	.12	0	.00
Obviously	20	.40	2	.03
Particularly	7	.14	1	.01
Partly	9	.18	0	.00
Perfectly	2	.04	0	.00
Possibly	16	.31	11	.16
Principally	5	.10	0	.00
Probably	20	.40	2	.03
Really	106	2.08	55	.79
Reasonably	4	.08	0	.00
Slightly	5	.10	0	.00
Strangely	4	.08	0	.00
Terribly	11	.22	0	.00
Total	395	7.76	117	1.68

the interviews, but it illustrates the considerable difference between the two groups. This difference would be even more striking were it not for the interview with Sinclair, who contributes just over half the tokens in the lower-class list. The difference, however, is not simply a matter of lexical choice (as Bernstein and his associates claim it is [e.g., Bernstein 1962; Lawton 1968]) but rather a difference in ways of communicating intensity.

The middle-class speakers are much more likely to use adverbs in −*ly* as a strong expression of the positive pole of intensity:

(16) a. IM107   I found it *extraordinarily* boring
      b. IM292   I got *absolutely* sick of
          293   doing nothing
      c. WG1484   but this zombie of a mother—*completely*
                     apathetic
      d. NM1381   oh well Jock it's *awfully* tasty
      e. DN152   a *terribly* crippled bent old woman

All the middle-class speakers use adverbs in −*ly* for this purpose and there are 25 clear examples in the middle-class interviews. There are only three lower-class examples of this use of adverbs in −*ly:*

(17) a. WR453   they were *spotlessly* clean
      b. EL3587   and it was an *exceptionally* good afternoon
      c. EL3275   she is an *exceptionally* clever little girl

The middle-class speakers also use adverbs in −*ly* in contexts where the effect is essentially redundant:

(18) a. WG102   and *suddenly* I'd made a breakthrough
      b. JM1875   and knocked myself out—*completely*
      c. DN1242   and then we *finally* got the length of a car
      d. NM1613   and this person used to come *regularly* to our
                     house

In the examples in (18) the essential meaning is conveyed by the verbs, and the adverbs somewhat redundantly reinforce this meaning. There are no obvious examples of this in the lower-class interviews.

As was shown in Table 9–2, the middle-class speakers also use *really* more than twice as frequently as the lower-class speakers. Labov observes:

> *Really* is one of the most frequent markers of intensity in colloquial conversation, and must figure large in any first approach to intensity in everyday conversation. It makes little contribution to cognitive or representational meaning, unless it is directly opposed to the unreal or the insincere. In fact, *really* can be

described as a "cognitive zero": it would have zero representational content in context-free information processing. (1984, 44)

Really, however, can intensify different constituents, with the possibility of a change of emphasis (though in many cases it is difficult to give a precise characterization of a difference in meaning). Stenström, in a study based mainly on the *London–Lund Corpus of Spoken English,*[1] observes, "What *really* does in the discourse is strongly related to its position, both within an utterance and in a sequence of interaction" (1987, 79). The least embedded position is when *really* occurs as sentential adverb, usually in clause-final position, as in the examples in (19):

(19) a. HG739   but you got nothing for it *really*
      b. AS3142 I would say my mother *really*
      c. NM1418 and *really* it was a gorgeous thing
      d. IM389   no I had no other hobbies *really*

Quirk and his colleagues call this use a "content disjunct" expressing a value judgment (1985, 622). Stenström suggests that *really* in final position has a "softening or cajoling effect" that is more personal and message-oriented than a discourse marker such as *you know* (1987, 72).

Really can also occur in truncated responses to questions:

(20)   (Did you ever think of going to live elsewhere?)
      AS2604   no not *really*

More than a third of the uses of *really* in the lower-class interviews are of the types illustrated in (19) and (20) compared with only 9% in the middle-class interviews. The majority of occurrences of *really* are modifying the predicate, as illustrated in (21):

(21) a. EL1231 and she *really* went to town on me
      b. AS1237 and I *really* enjoyed that
      c. AM1279 and you *really* learned the stuff
      d. JM1130 I *really* liked Ayr very much

Where there is an auxiliary, *really* normally follows the first auxiliary:

(22) a. EL141   they would *really* be on the carpet
      b. AS3105 it should *really* start from the home
      c. AM1203 where he was *really* making fun of people

[1]*The London-Lund Corpus of Spoken English* is a collection of approximately 500,000 words recorded in a variety of situations, conversations, broadcasts, and public speeches (Svartvik and Quirk 1980).

> d. WL2150   no a thing
>   2151   I've *really* enjoyed awful much

This is generally, but not always, the case when there is a negative, whether there is an auxiliary or not:

(23) a. NM445   I didn't *really* go to Sunday school
      b. JM557   the girls didn't *really* have any game for the summer
      c. IM545   I never *really* gave it a thought

There are 44 examples of *really* with negative predicates. In 31 cases (70%) the negative precedes *really,* and in the remaining 13 (30%) *really* precedes the negative. These latter examples suggest that the placement of *really* has a semantic value. There are 8 examples (62%) with the verb *know,* all but one of them in the form *I really don't know,* and the other *he really didn't know.* It seems as if in the absence of any quantifier, knowing is not seen as gradable. The intensifier thus modifies the higher predicate with a possible paraphrase, "It is really the case that I do not know." The one example in which *really* follows the negative in a predicate with *know* is (24):

(24)   AS429   I didn't *really* know much aboot it at the time

This remark comes at the end of a narrative in which Sinclair explains how he lost his first job because of a dispute between his parents and the farmer over compensation for an injury. Sinclair clearly knew something about it, but not much. A possible paraphrase is, "It was not the case that I knew really much about it." Another example where the placement of *really* may affect its scope is (25):

(25)   NM1273   Hogmanay *really* didn't affect the young members of the family

Here the intensifier indicates that the young members of the family were not affected in any way, whereas the placement of *really* after the negative would allow that possibility that they were affected to some extent, as in an example from MacDougall (AM1079 although it didn't really affect me an awful lot).

Really also usually follows the copula:

(26) a. EL134   but I was *really* the granddaughter
      b. MR707   that was *really* a time
         708   that I enjoyed
      c. AM924   and nothing else was *really* all that important
      d. JM579   it's *really* an awful shame

In such cases *really* intensifies the whole predicate; but sometimes (particularly where an adjective follows the copula) it is impossible tell whether the intensification is of the predicate or of the adjective itself:

(27)　a. AS2191　　he's *really* terrific
　　　　b. JM612　　　that was *really* good

(27a) could be paraphrased, "It is really the case that he is terrific" or "It is the case that he is really terrific." Again, the placement of *really* before the copula eliminates the ambiguity:

(28)　a. NM089　　it *really* was a good school
　　　　b. JM576　　　it *really* was something

There are also clear cases where the adjective is directly intensified:

(29)　a. AS1835　　we've got the name of
　　　　　　1836　　　being clannish but not *really* clannish
　　　　b. AM1695　because I—I find it *really* helpful
　　　　c. AM377　　he gave—a *really* excellent lecture course
　　　　d. NM1059　he would have been *really* happier

(29d) is the only example where *really* does not directly follow the modal auxiliary in an affirmative clause, and this suggests that it is intensifying the adjective.

There are 95 examples of *really* with an auxiliary, copula, or negative. In 72 cases (76%) *really* occurs second; in the remaining 23 (24%) *really* precedes. In the cases where *really* precedes one of these items, it is usually possible to find a plausible explanation. This suggests that the unmarked position for *really* within a clause is before the predicate but that in certain cases a position before one of these operators is preferred.

Finally, there are four examples in the middle-class interviews of *really* embedded in an infinitive:

(30)　a. WG565　　and it ill becomes us
　　　　　　　　　　　to be *really* squabbling and fighting amongst ourselves
　　　　b. AM1411　then the local people are going to *really* not have an awful lot of safeguards
　　　　c. AM1590　which we tend
　　　　　　1591　　　not to *really* get in the formal teaching at Glasgow
　　　　d. IM490　　I didn't have the time
　　　　　　491　　　to *really* train

Examples of this kind do not occur at all in the lower-class interviews, and clear examples of *really* intensifying a constituent other than the predicate or the whole clause are rare. Given that the proportion of clause-external *really* is three times greater in the lower-class interviews, the freedom to use *really* as a modifier of a wide range of constituents seems to be mainly a middle-class feature. In fact, only two of the lower-class speakers, Laidlaw and Sinclair, use *really* with any freedom; and between them they account for 75% of the examples in the lower-class interviews. Among the middle-class group Mac-Dougall, the most highly educated speaker, accounts for a large proportion of the examples (47%); while Nicoll, whose speech sometimes shows similarities to the lower-class group, has (surprisingly) only a single example of *really* in an interview of 15,200 words.

Lang and Nicoll are the most fluent speakers in their groups, and their interviews contain the highest proportion of narratives; but Lang has only three examples of *really,* and Nicoll only one. This is a frequency of .13 per thousand words, compared with MacDougall's 4.5 per thousand words. There is no way in which MacDougall's interview could be considered a better example of colloquial speech than Lang's or Nicoll's. Thus, whatever motivates the use of *really,* it does not appear to be simple conversational exuberance. Rather, the more cautious and better-educated speakers use it most frequently. On the other hand, Stenström (1987, 67) points out that *really* is much more frequent in speech than in writing.

It is not only adverbs in *−ly* that the middle-class speakers use more frequently to express positive intensity; they also use *very* more than three times as frequently as the lower-class speakers. *Very* can be used as an intensifier with adjectives, adverbs, and quantifiers, as shown in (31a–c):

(31) a. EL440   Belgian girls are *very* funny
      b. JM583   *very* definitely a great pride in the school
      c. HG514   *very* few of them kent one another anyway

There is, however, no difference in the distribution of these three uses between the two social class groups, as shown in Table 9–3. Within the groups, the picture is less clear. Lang has only 4 examples (accounting for 7% of the lower-class examples), and Sinclair has 22 (39%). In the middle-class group, Nicoll has 41 examples (27%), and MacDougall only 9 (6%). The pattern of use of *very* is thus much different from that of *really.*

Another intensifier that the middle-class speakers use more frequently than the lower-class speakers is *quite.* There are 54 examples in the lower-class interviews, .79 per thousand words. In the middle-class interviews there are 116 examples, 2.27 per thousand words. Once again, Sinclair is responsible for a high proportion (46%) of the lower-class examples. Without Sinclair's

Table 9.3. Use of *Very* As an Intensifier

Class	W/Adjectives		W/Adverbs		W/Quantifiers		/1,000 wds.	n
	%	n	%	n	%	n		
Lower	65	37	17.5	10	17.5	10	.82	57
Middle	68	105	16	25	16	24	3.03	154

interview, the frequency in the lower-class interviews is only .62 per thousand words.

Quite can modify an adjective, an adverb, or a noun phrase, as shown in (16):

(32)   a. WL594   he was *quite* good
       b. WG1111   I found it *quite* traumatic
       c. AS017   visiting them *quite* regularly you know
       d. AM003   but that was *quite* recently
       e. EL152   he's *quite* a lad
       f. JM2435   and I had *quite* a reputation

The frequency of each of these uses of *quite* differs between the two groups, as shown in Table 9–4. The lower-class speakers use *quite* twice as often before a noun phrase and only half as often before an adverb. In fact, the two lower-class examples of *quite* before an adverb come from a single interview—Sinclair's. As in the case of *really,* MacDougall is responsible for the largest number of middle-class examples (32%) of *quite,* while Nicoll accounts for only 12%. MacDougall also accounts for almost half (46%) of the examples modifying an adverb.

Labov (1984, 45) points out that *just* can be both an intensifier and a minimizer. The most detailed accounts of the meaning and function of *just* are those in Lee (1987) and Quirk and his colleagues (1985). This is one item that is used equally frequently by the middle-class speakers (4.6 per thousand words) and the lower-class speakers (4.7 per thousand words); and Sinclair's

Table 9.4. Use of *Quite* As an Intensifier

Class	Adjective		Adverb		Noun Phrase	
	%	n	%	n	%	n
Lower	41	22	4	2	56	30
Middle	64	74	11	13	25	29

figures are consistent with those of the other lower-class speakers. Thus, there seems to be little social class significance to this item. Although in individual cases the meaning is often hard to pin down exactly, the evidence from the Ayr interviews confirms the findings of Lee (1987) that four uses can be distinguished. The first is close association with a verb in the sense of "not long ago," as shown in (33):

(33) a. HG051   I was *just* saying to John MacGregor
      b. AS2581   and I *just* had come out
      c. JM1911   a friend of mine had *just* got a new Baby Austin
                  with a sunshine roof
      d. DN2167   we *just* celebrated our forty-first anniversary

This is the least frequent use (less than 3%) in both sets of interviews. Quirk and his colleagues (1985, 194) describe this use as a *time adverbial* emphasizing recency; but Lee suggests that this is only one aspect of a more general category, which he calls *specificatory meaning* (1987, 388). This category includes the third-most-frequent use of *just* in the Ayr interviews, that is, as an intensifier with the general sense of "exactly" (27% overall):

(34) a. WL018   there *just* next to Prestwick airport
      b. AS112   but—eh—*just* like everything else
      c. NM775   this was *just* before the war
      d. JM1412   and we both stayed *just* by the tennis courts

This use is most common with adverbial expressions of time and place.

The second-most-frequent use (30% overall) is in the sense of "only" or "no more than":

(35) a. HG1560   it was *just* bare rock
      b. WL1668   it was *just* a wee cub you see
      c. JM960   actually it isn't *just* my own opinion
      d. DN1827   he was *just* a barefooted boy

Quirk and colleagues (1985, 604) and Lee (1987, 384) label this the "restrictive meaning" of *just,* though (as Lee points out [1987, 387]) their categories are not identical. This use overlaps with one that can perhaps best be paraphrased as "simply":

(36) a. WL735   we *just* lifted the streetcher
      b. MR686   it *just* fell away
      c. JM942   then you *just* quietly drop them from the team
      d. NM271   but we *just* termed it our society—our secret
                  society

This is the most frequent use (41% overall) and probably can be taken to be the neutral, or unmarked, meaning of *just*. Quirk and his colleagues label this use a *down-toner,* specifically a "diminisher" (1985, 598). Lee calls it the "depreciatory meaning." It is perhaps significant that while *just* in the sense of "only" can follow a negative, as in 35c, when *just* precedes the negative it can never have this interpretation:

(37) a. IM026   I *just* (= simply/*only) can't remember the name of it now
      b. AS1861   it *just* (= simply/*only) wouldnae go down at all

The lower-class speakers, however, sometimes use *just* in a location that might give a confusing idea of its scope:

(38) a. WL1820   he *just* died last year
      b. WL1829   I was *just* looking at them the other day
      c. AS987   and *just* I could get my head shoulders out through it
      d. AS1662   at that time he *just* was an ordinary plumber

In (38a), for example, it is obvious that the scope is "just last year" and not "just died."

In his conclusion, Lee observes:

> Although it is possible to identify different categories of meaning for *just* in different utterances, these categories are linked to each other in intricate ways. Examples can be found in which two (or even more) meanings combine, so that one type of meaning overlays and shades into another. Borderline cases can be identified where it is difficult to decide to which category a particular case should be assigned. (1987, 395)

Lee argues that his examination of *just* supports the views of Moore and Carling (1982), where meaning emerges from a complex interaction of various factors rather than being located primarily in individual lexical items or structures. Moore and Carling place the emphasis not on language as an independent entity but rather as "an epiphenomenon on the accumulated and generalised experience of its users" (p. 216). Their most fundamental assumption is: "that meaning should not be viewed as an inherent property of words, but as an emergent property of utterances" (p. 211). Lee claims that *just* is a very good illustration of this emergent view of meaning: "The fundamental question has to do with the nature of the interpretive principles which human beings bring to the process of understanding language. It is clear that these principles are no less complex in the case of an element as unobtrusive as *just* than those which apply to more obviously problematic cases" (1987, 397).

Table 9.5. Evidential Adverbs

Adverb	Middle-class		Lower-class	
	n	/1,000 wds.	n	/1,000 wds.
Basically	9	.18	1	.01
Certainly	14	.28	10	.14
Exactly	8	.16	1	.01
Generally	9	.18	3	.04
Literally	5	.10	0	.00
Maybe	66	1.30	102	1.46
Normally	4	.08	2	.03
Particularly	7	.14	1	.01
Perhaps	16	.31	1	.01
Possibly	13	.26	11	.16
Primarily	1	.02	0	.00
Probably	20	.39	2	.03
Obviously	20	.39	2	.03
Seemingly	1	.02	2	.03
Surely	0	.00	2	.03
Total	193	3.79	140	2.01

Lee drew his examples from a number of doctor–patient interviews in the Brisbane area. It is remarkable that his results should be so similar to those based on the Ayr interviews, suggesting that the complexity of the unobtrusive item *just* has a very stable basis.

Finally, the use of adverbs as evidentials (cf. Chafe 1987, 261–72) is almost twice as frequent in the middle-class interviews (3.8 per thousand words, compared with 2.1 per thousand words in the lower-class interviews), as Table 9–5 shows. Most of these adverbs come into the category Chafe calls "degrees of reliability" (1986b, 264). In his comparison of conversational English with academic writing Chafe found that expressions of this type were more than twice as frequent in academic writing. If his findings prove to be characteristic of a difference between written and spoken language in general, once again, the middle-class speakers in Ayr will have shown a feature that is closer to the written norm than is found in the lower-class interviews.

Biber and Finegan (1988), on the other hand, classify adverbials in six categories as "markers of stance" and identify eight types of "stance" according to the frequency of categories of adverbials used. They characterize one style as "faceless" because of the absence of adverbials that "overtly mark attitudes or commitment towards the message" (p. 22). The majority of texts in this style in their sample are written, and none of the spoken texts are conversational. The speakers in the London–Lund Corpus of Spoken English, from which Biber and Finegan took their texts of speech, were "mostly middle-class, university-educated adults" (p. 5); and this may account for the relatively high frequency of "stance adverbials" in their spoken texts. The much lower frequency of such adverbials in the lower-class Ayr interviews underlines the need for a more representative corpus of spoken English to avoid the inherent class and educational bias of the London–Lund corpus.[2]

Other Forms of Intensification

Labov (1984) deals primarily with universal quantifiers, and he points out that they are frequently used as intensifiers when it is clear that they are not intended to be interpreted literally. He calls such a use "loose interpretation." Examples of universal quantifiers that require a "loose interpretation" in the Ayr interviews are given in (39):

(39) a. MR391   I mean I've been *all* over England and Wales and Ireland

[2]One curious feature of the evidential adverbs is the preference for *maybe* over *perhaps*. *Maybe* occurs with a frequency of 1.46 per thousand words in the lower-class interviews and 1.30 in the middle-class interviews. There is only 1 occurrence of *perhaps* in the lower-class interviews (from Laidlaw), a frequency of .01, and only 16 in the middle-class interviews, a frequency of 0.31. This is the reverse of the situation in most written English. The Lancaster–Oslo–Bergen corpus of written British English, for example, contains 85 examples of *maybe* and 407 examples of *perhaps* (Johansson and Hofland 1989). I also consulted a number of concordances of British and American writers:

	Maybe	Perhaps
Blake	0	61
Browning	12	131
Byron	6	286
Conrad	43	806
Dryden	2	77
Poe	1	201
Shaw	6	478
Wordsworth	0	81

There were, however, three exceptions to the general pattern:

	Maybe	Perhaps
Burns	4	0
Langston Hughes	27	11
Yeats	45	2

The use of forms such as *maybe* might shed some light on the historical development of spoken English.

   b. WR750   well *all* the villages had juvenile teams in them
               days ken
   c. WL2054   of coorse nooadays it's *every* hoose
      2055   has got a television sort of thing
   d. WG1152   and they also had the TV on *all* the time
   e. NM1622   but they certainly got *all* our clothes
   f. DN22   my wife polished it *every* morning in life

Although Labov does not claim that his analysis is intended to be a sociolinguistic study of intensification and quantifiers, he points out that A.P., a
middle-class lawyer in Philadelphia, used universal quantifiers much less
frequently than the two working-class men whose interviews he had analyzed.
This pattern is repeated in Ayr. The lower-class speakers use universal quantifiers with a frequency of 13.9 per thousand words, compared with 7.1 per
thousand words in the middle-class interviews. This is the reverse of the
situation with adverbs in −*ly,* where it is the middle-class speakers who show
the more frequent use. If universal quantifiers and adverbs are alternative
ways of signaling intensity, then the choice of one or the other is largely a
matter of style, since both groups of speakers use both devices.

One method of intensification, however, appears to be restricted to one
social class group. The lower-class speakers often use repetition for emphasis,
as in the examples in (40):

(40)   a. WL857   you ken I saved that man's haund
          858   I went up the pit with him
          859   and I lost time
          860   for going up the pit with him
          861   I come hame
          862   you see the ambulance come to Ayr County
          863   and I was coming to Whitletts here you see
          864   I stayed in twenty-three at that time
          865   so my time was cut
          866   for coming to the hospital with him
          867   MY TIME WAS CUT FOR THAT
       b. EL706   if I made a mistake
          707   or if I was reported for anything
          708   I got a thumping
          709   and I mean I GOT A THUMPING
          710   it wasnae just a clip on the ear and sent ootside

Tannen has argued that repetition is an important part of conversational coherence: "By facilitating production, comprehension, connection, and interaction in these and other ways, repetition serves an over-arching purpose of

creating interpersonal involvement" (1989, 52). In the repetitions for emphasis, such as those illustrated in Item 40, the tempo is slower and the voice
slightly louder. There are several examples of this in the lower-class interviews but none in the middle-class interviews. The middle-class speakers, on
the other hand, sometimes use breathy voice for emphasis:

> (41)   a. IM767   I remember even in my time
> 768   people used to come into the shop and say
> 769   "Oh it's so nice here
> 770   it's not like this in England you know"
> 771   the inference being
> 772   that they were well attended to
> 773   but *not by God now* [breathy voice]
> b. DN153   and as she passed me
> 154   she said
> 155   "Goodnight Your Highness"
> 156   [nods slightly] just her head
> 157   she couldnae move—*just her head* [breathy voice]
> 158   it was wonderful
> 169   it made the day

There are too few examples to be sure that this is not simply an idiosyncratic
feature, but three of the middle-class speakers use breathy voice for emphasis.

Conclusion

I have shown that there are noticeable differences between the lower-class and
middle-class speakers in how they signal intensity. The lower-class speakers
make more use of differences in word order, universal quantifiers, and repetition. The middle-class speakers make much more use of a wide range of
adverbs. There are also possible differences in prosodic features such as
emphatic stress or breathy voice.

Since the two social class groups apparently employ different discourse
strategies, it is legitimate to ask whether these strategies can be characterized
in general terms. Haiman (1985, 166), following Saussure (1922, 183), suggests that languages with a more restricted vocabulary are more likely to make
greater use of "grammatical" constructions. Although—as was shown in
Chapter 8—the vocabulary of the lower-class speakers is by no means im-
poverished, it is slightly less varied than that in the middle-class interviews;
and it is striking that the lower-class use of syntactic dislocation to indicate
intensity should be almost in complementary distribution with the use of

intensifying adverbs in the middle-class interviews. Halliday (1987, 71) has argued that spoken language tends to favor greater "grammatical intricacy," while written language favors greater "lexical density." It is hardly surprising that lower-class speakers, with less education and probably less recourse to written materials in their daily lives, should employ discourse strategies that are characteristic of spoken language or that middle-class speakers should be more "bookish" in their speech. (Though the greater frequency of *really* in the middle-class interviews cannot be attributed to the influence of the written language.) These differences are another illustration relevant to Hymes's caveat regarding the dangers of polarizing dichotomies (1986, 50).

Because the examples were taken from a small number of speakers in single dyadic interviews, it is impossible to know to what extent the same stylistic choices will be made in other situations with other participants, but the analysis illustrates the kind of alternatives available to members of the speech community. Although the analysis has been presented in terms of social class differences, it must be emphasized that there may be other factors operating that are just as important. For example, there are more narratives in the lower-class interviews. Future investigation will explore the contrast between the language used in narrative sections and the language of the rest of the interview. It is notable, however, that where quantitative and qualitative methods of analyzing connected discourse have been employed (e.g., Labov 1984; Lee 1987; Schiffrin 1987), the results are generally consistent with those that have been found in the Ayr study. This suggests that there may be important characteristics of colloquial speech that are not narrowly localized. This will make Hymes's goal in the quotation cited at the beginning of this chapter a more manageable one.

10

Discourse Features

Of all the aspects of language examined in the present work, discourse features are the ones that are most likely to be constrained by the interview situation. For the most part, the interviews are dyadic exchanges even when more than two people are present; and they are unbalanced exchanges because the role of the interviewer is to yield the floor whenever possible. Thus, the dynamics of turn taking (Sacks, Schegloff, and Jefferson 1974; Duncan and Fiske 1985) are reduced to a minimum. There are also differences in genre and topic that result from different responses to the opportunities offered in the interview. This makes quantitative analysis particularly tricky; but there may be some value in presenting the raw figures—if mainly for the purpose of providing a baseline for comparison with other studies. As was pointed out in Chapter 2, the frequency of the interviewer's questions and the length of the responses vary considerably. Some interviews contain several "performed narratives" (Wolfson 1978) with quoted direct speech, asides, expressive sounds and gestures, and frequently the use of the so-called historical present tense (Wolfson 1978; Schiffrin 1981); while others have no narratives at all and little or no quoted direct speech. The proportion of quoted direct speech for each speaker is shown in Table 10–1. There is a similarity between the proportion of quoted direct speech and the number of narratives, as can be seen in Table 10–2. These figures do not include what Horvath (1987) has called "reminiscences" or "anecdotes" where the action is less specific or localized. Since the narratives vary greatly in length, a more significant figure is perhaps the proportion of the interview devoted to narrative. The figures in Table 10–3 are very approximate, because it is often difficult to determine at which point a narrative actually begins or ends. Gross differences of this kind obviously affect the frequency of certain features. For these reasons, quantitative methods must be used with caution in dealing with discourse features,

138

Table 10.1. Quoted Direct Speech

Respondent	Percentage
Lower-class	
Laidlaw	11.5
Lang	10.0
Rae	7.5
Sinclair	3.6
Ritchie	3.4
Gemmill	2.7
All LC	6.8
Middle-class	
Nicoll	16.2
Gibson	10.4
MacGregor	4.2
Muir	.9
MacDougall	.4
Menzies	.2
All MC	7.7

Table 10.2. Narratives

Respondent	Number of Narratives
Lower-class	
Lang	24
Laidlaw	14
Sinclair	13
Gemmill	10
Rae	6
Ritchie	2
All LC	71
Middle-class	
Nicoll	15
MacGregor	9
Gibson	6
MacDougall	1
Muir	0
Menzies	0
All MC	31

Table 10.3. Narrative Portion
of Interview

Respondent	Percentage
Lower-class	
Lang	50
Laidlaw	30
Sinclair	25
Rae	20
Gemmill	15
Ritchie	5
All LC	29
Middle-class	
Nicoll	45
MacGregor	25
Gibson	25
MacDougall	2
Muir	0
Menzies	0
All MC	22

though Schiffrin (1981, 1983, 1987) has shown how informative such methods can be.

The first features to be examined are what have been called "pragmatic particles" (Östman 1982) or "discourse markers" (Schiffrin 1987). Östman gives a preliminary characterization of the class:

> Typically, a pragmatic particle would be (a) short, and (b) prosodically subordinated to another word. It would (c) resist clear lexical specification and be propositionally empty (i.e., it would not be part of the propositional content of the sentence). Furthermore, it would (d) tend to occur in some sense cut off from, or on a higher level than, the rest of the utterance, at the same time as it tends to modify that utterance as a whole. (1982, 149)

Östman admits that there are problems in trying to apply these criteria categorically in delimiting the class of pragmatic particles. Schiffrin sees the problem as one of overlapping categories and fuzzy boundaries: "Standing stubbornly in the middle of both sentences and discourse grammars, and semantics and pragmatics, are linguistic elements like *and, but, so, because, like, now, I mean, y'know, well, why, oh,* and *anyway*" (1983, 1). This list includes items such as *and, but, because,* and so on, which have a clear syntactic function. Although Schiffrin is correct in pointing out that such items have a pragmatic function as well as a syntactic function, this distinguishes them from most elements in the sentence only in degree, not in kind. The true pragmatic particles, in Östman's sense, have no clear syntactic function or semantic interpretation. The most obvious example is *oh*, which has a discourse function that can be described fairly explicitly, although its syntactic and semantic functions are harder to specify.

The definition given above is taken from an unpublished paper by Schiffrin, and she would not necessarily choose to formulate it in this way now; but I have quoted it because it was influential in the analysis of discourse markers in the Ayr interviews, which was carried out before her book, *Discourse Markers* (1987), appeared. In the book, Schiffrin operationally defines discourse markers as "*sequentially dependent* elements which bracket units of talk" (1987, 31, emphasis original) and goes on to conclude that discourse markers help to produce coherent discourse by allowing speakers "to construct and integrate multiple planes and dimensions of an emergent reality" (p. 330).

The set of discourse markers to be examined consists of *oh, well, (you) see, I mean, you know, (you) ken, (of) course, anyway, in fact, now, here,* and *mind you.* Examples to illustrate these discourse markers are given in (1):

(1) a. WG205   *oh* I've given it up now
        MR811   *oh* Annbank havenae a twang
     b. NM784   *well* I was needed at home
        EL1421   *well* yon two couldnae care less
     c. DN2207   it was ridiculous *you see*
        WL065   and I filled the cairts *you see* at the station
     d. AM558   *I mean* nobody came out
        AS895   *I mean* they were steel
     e. IM120   and there was nobody *you know* in line for it
        WL442   and *you know* kenning him that weel
     f. WL1646   if they were daeing ocht *you ken*
        WR811   I say them baith *ken* at Burns Suppers
     g. JM846   and he *of course* was first choice
        WL2327   *coorse* the roads is all altered noo tae
     h. AM1430   *anyway* it doesn't matter
        HG819   och there nae tatties noo *anyway*
     i. WG330   *in fact* one wee chap got nightmares
        AS800   he's the auldest in the family *in fact*
     j. DN2267   *now* I'm not a roll man a croissant man
        WL2119   *now* you're to behave yourselves lassies
     k. DN1254   and *here* this comes in
        HG174   *here* the blooming war feenished again

1. DN2748   *mind you* the smell would go through too
   EL3219   we'd never been in the shop *mind you*

As in any aspect of language, it is necessary to distinguish superficially similar forms that have a different function, as the examples in (2):

(2) a. EL3583   he was *well* liked
        MR008   no she hasn't been keeping *well* at all
     b. AS2208   you'll find oot
        2209   what *I mean*
        WG074   by that *I mean*
        075   speaking—
        076   reading aloud
     c. DN500   whether *you know* it or not
        AS2304   not what *you know*
        2305   who you knew
     d. WG165   as there is *now*
        AS598   many a time *noo* I wished
        599   I had

In the examples in (2) *well, I mean, you know,* and *now/noo* are not discourse markers because they are directly linked to the syntax of the clause in which they occur and have a totally different function from the homonymic discourse markers. The equivalent items in (1) could be omitted without disturbing either the syntax or the prepositional content of the clauses. Only items of the type illustrated in (1) will be tabulated as discourse markers.

Discourse markers are more than twice as numerous and frequent in the lower-class interviews as in the middle-class interviews. Table 10–4 shows the number of these discourse markers for each speaker, the frequency per thousand words, and the percentage of syntactic units that contain a discourse marker. There is no doubt that these discourse markers are much more frequent in the interviews with the lower-class speakers; and although there is overlap, the highest individual figures in the middle-class group are below the average for the lower-class group. The rank order within each group, however, is perhaps surprising. As was pointed out in Chapter 2, Ritchie, Muir, and Rae were not the most forthcoming of respondents; yet Rae and Muir have the highest frequencies of discourse markers in their respective groups, and Ritchie uses them with a higher frequency than all but three of the speakers. Östman has argued that "the occurrence of pragmatic particles in a discourse turns out to be a sufficient condition for regarding that discourse as having a high degree of impromptuness" (1982, 170). Östman also characterizes impromptu speech as "spontaneous, everyday face-to-face interaction" that focuses "more on the cognitive and interactional processes in-

Table 10.4. Number and Frequency of Discourse Markers

Respondent	n	/1,000 wds.	% syntactic units with discourse markers
Lower-class			
Rae	317	61	43
Lang	532	33	23
Sinclair	608	29	19
Ritchie	129	31	17
Gemmill	180	20	11
Laidlaw	204	16	8
All LC	1,970	29	18
Middle-class			
Muir	117	27	14
Menzies	90	19	11
Nicoll	222	15	8
MacDougall	106	10	7
Gibson	86	10	6
MacGregor	69	9	6
All MC	690	14	8
Total	2,660	22	14

volved, than on the ultimate linguistic product" (p. 154). The keywords here are probably *interactional* and *everyday*. As was shown in Table 2–2, the highest frequency of interviewer's questions occurs in the interviews with Ritchie, Rae, Menzies, and Muir. In the interviews with Rae and Ritchie more than half of the questions have a response that includes *oh* or *well;* in the interviews with Muir and Menzies the proportion is more than a quarter. These are also the shortest interviews, so that the high number of such responses increases the frequency. These were the four interviews in which the interviewer was working hardest to obtain a response. In that sense, they are more "interactional" than the others.

The other keyword is *everyday*. There are few narratives in these four interviews. There is a sense in which Lang, Laidlaw, Nicoll, and Sinclair are putting on a "performance" for the interviewer's benefit. Although discourse markers are not absent from narratives, their role is not necessarily the same as in "everyday face-to-face interaction." For example, only *oh, well, of course,* and *now* occur in quoted direct speech. Since Nicoll, Laidlaw, and Gibson have the highest proportion of quoted direct speech, this may contribute to the relatively low frequency of discourse markers in their interviews. It would not, however, account for the high frequency of discourse markers in

Lang's interview, since he has the next-highest proportion of quoted direct speech. As will become apparent in the more detailed analysis, there are also idiosyncratic factors that affect the use of discourse markers; but even at this level of analysis it is clear that the problems involved in characterizing style are immense (cf. Traugott and Romaine 1985). Table 10–5 gives the figures for each discourse marker. It can be seen from Table 10–5 that only two of these discourse markers are used proportionally more frequently by the middle-class speakers: *of course* and *now*. In the use of *you know* the frequency is roughly the same in both social class groups; but for all the other discourse markers the frequency is greater in the lower-class interviews, including *(you) ken,* which is used exclusively by lower-class speakers. Table 10–6 shows the rank order of discourse marker use for each social class group. In the lower-class interviews *oh* is more frequent (7.2 instances per thousand words) than *well* (5.4 per thousand words). In the middle-class interviews the order is reversed, *oh* (2.7 per thousand words) and *well* (3.7 per thousand words). It is perhaps significant that there are few examples of *oh* in MacDougall's inter-

Table 10.5. Distribution of Discourse Markers

Respondent	O	W	S	I	Y	K	C	A	F	N	H	M	Total
Lower-class												
Rae	93	34	6	4	2	158	4	10	1	0	4	1	317
Lang	154	79	124	34	59	35	23	2	13	2	3	4	532
Sinclair	51	151	106	185	48	7	5	10	24	16	4	1	608
Ritchie	54	18	6	9	30	0	9	0	2	0	0	1	129
Gemmill	74	79	4	0	4	1	5	25	3	1	15	1	180
Laidlaw	65	43	15	40	11	1	4	7	3	12	2	1	204
All LC	491	372	261	272	154	202	50	54	46	31	28	9	1,970
Middle-class												
Muir	11	24	1	14	33	0	18	11	4	1	0	0	117
Menzies	20	24	2	11	21	0	5	1	0	5	0	1	90
Nicoll	55	42	70	3	12	0	15	0	1	20	2	2	222
MacDougall	8	51	0	8	27	0	4	1	7	0	0	0	106
Gibson	28	27	4	4	7	0	5	2	3	6	0	0	86
MacGregor	15	20	3	2	7	0	13	2	4	3	0	0	69
All MC	137	188	80	42	107	0	60	17	19	35	2	3	690
Overall	628	560	341	314	261	202	110	71	65	66	30	12	2,660

Note: O = *oh*, W = *well*, S = *(you) see*, I = *I mean*, Y = *you know*, K = *(you) ken*, C = *(of) course*, A = *anyway*, F = *in fact*, N = *now*, H = *here*, M = *mind you*.

Table 10.6. Rank Order of
Discourse Markers

Lower-class	Middle-class
oh	well
well	oh
I mean	you know
(you) see	(you) see
(you) ken	I mean
you know	(of) course
anyway	now
(of) course	in fact
in fact	anyway
now	mind you
here	here
mind you	

view. As pointed out earlier, MacDougall is by far the best-educated speaker in the sample; but the low frequency of *oh* may be simply an idiosyncratic feature of his speech or a reflection of the way he dealt with academic topics in the interview.

The use of the verbal group of discourse markers (*I mean, (you) see, (you) ken, you know*) shows the greatest social class difference. The overall frequency of these four items in the lower-class interviews is 13.1 per thousand words compared with only 4.5 per thousand in the middle-class interviews. This difference would be even greater were it not for the high frequency of *(you) see* in Nicoll's interview. Nicoll is the middle-class speaker whose interview occasionally shows features of lower-class speech (e.g., the use of the negative clitic —*nae*), so it is interesting that he is responsible for the great majority (88%) of examples of *(you) see* in the middle-class interviews. In the other middle-class interviews the frequency of *(you) see* is only .3 per thousand words, compared with 3.8 per thousand words in the lower-class interviews and 4.6 per thousand words in Nicoll's interview. It seems reasonable, therefore, to conclude that *(you) see*, like *(you) ken*, is primarily a feature of lower-class speech, though not quite as salient. *You know*, on the other hand, shows no such difference, since it is almost as frequent in the middle-class interviews (2.1 per thousand words) as in the lower-class interviews (2.3 per thousand words). This contradicts the views of Bernstein (1971) and Huspek, who assume that *you know* is characteristic of "socially disadvantaged speakers" (Huspek 1989, 664).

There is, however, great individual variation in the frequency with which each of this group of discourse markers is used. The extreme example is Rae, who uses *(you) ken* in 21% of the syntactic units. The effect which this frequency can have is illustrated in (3):

(3)   623   oh but this I dae remember
         624   the pits were privately owned
         625   and aw the miners stood at the corner *ken*
         626   they backed their wee lines
         627   the bookie was standing there and aw
         628   sixpenny doubles and one thing and another *ken*
         629   here this day I wondered
         630   what was wrang
         631   aw the auld yins took their caps off *ken*
         632   because it was the pit owner
         633   that was going by *ken*
         634   for respect *ken*
         635   an auld auld man he was *ken*
         636   come by in a chauffeur-driven motor *ken*
         637   the old yins daffing the cap
         636   in oor days they'd be chucking stanes at that motor

Rae is responsible for 78% of the examples of *(you) ken* in the interviews. Similarly, Sinclair provides 68% of the examples of *I mean* in the lower-class interviews; and he and Lang account for 88% of the instances of *(you) see* in the lower-class group. Gemmill, on the other hand, has the lowest frequency (1.0 per thousand) of these verbal discourse markers in the whole sample. Differences as great as these are not solely governed by topic or genre but must reflect some personal preferences in stylistic choice. Other examples of this will be illustrated later.

The use of *here* and *mind you* is predominantly in the lower-class interviews. The remaining four discourse markers, *(of) course, anyway, in fact*, and *now*, taken as a group, are used with the same frequency by both social class groups (2.7 per thousand words in the lower-class interviews, 2.6 per thousand words in the middle-class interviews).

I shall now consider the function of the discourse markers.

Well and Oh

Since *well* and *oh* have many similarities, it is helpful to consider them together, although they are far from identical in their use. *Well* has frequently been discussed in conversational analysis and in discourse analysis, the most

Table 10.7. *Well* in Answers to Questions

Use of *Well*	Yes/no questions		WH-questions	
	%	n	%	n
Lower-class				
With *well*	8	15	35	36
Without *well*	92	169	65	67
Middle-class				
With *well*	14	23	40	30
Without *well*	86	147	60	45
All				
With *Well*	11	38	37	66
Without *Well*	89	316	63	112

recent being by Schiffrin (1985, 1987), who also summarizes the earlier work. After examining various uses of *well* Schiffrin concludes: "*Well* is one device used by speakers in their attempts to build coherence: it anchors the speaker in a conversation precisely at those points where upcoming coherence is not guaranteed" (1985, 662). One of these points is the answer to a question. Schiffrin shows that in her corpus *well* is significantly more likely in response to WH-questions than to *yes/no* questions. This is confirmed by the Ayr interviews, as can be seen in Table 10–7. Schiffrin argues that the reason is that *yes/no* questions constrain the answer more narrowly: "When the conditions for the upcoming coherence of an answer have been relatively delimited by the form of the prior question, *well* is not as useful for marking the answer as a coherent response" (1985, 644). This, too, is borne out by the Ayr interviews. It is, however, not only *yes/no* questions that constrain the response narrowly. Many of the initial questions in the interviews are very specific WH-questions (e.g., *When were you born? What did your father do?*). There are 25 such questions in the first two pages of each transcript, and 23 (92%) of the responses do not begin with *well* but instead provide a direct response to the question. As Schiffrin points out, *well* is used when the speaker is giving a response that is not a direct answer to the question. Some examples are given in (4):

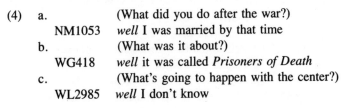

(4) a.   (What did you do after the war?)
       NM1053   *well* I was married by that time
    b.   (What was it about?)
       WG418   *well* it was called *Prisoners of Death*
    c.   (What's going to happen with the center?)
       WL2985   *well* I don't know

d.   (What was it like when you were growing up?)
EL179   *well* it differs from one—perhaps—eh—
180   I wasn't brought up—eh—easy

In the examples in (4) the speakers are not really providing the information requested by *what,* although they are responding in a positive way to the question.

One of the other uses of *well*—also noted by Schiffrin—is *self-repair.* This is more frequent in the middle-class interviews. Of 188 instances of *well* in the middle-class interviews, 47 (25%) can be classified as examples of self-repair, compared with only 5% (20/372) in the lower-class interviews; but this is probably because self-repair is more common in the middle-class interviews. Examples are shown in (5):

(5) a. JM600   the jan—the old janitor—*well* actually it was the
                        son
      b. AM1579   for the next four weeks—*well* for the next two
      c. WL1751   and he'd this little kind of—*well* an auld-fashioned
                        caur anyway
      d. EL3876   a black man—*well* I shouldnae say a black man—
                        a colored man

Schiffrin also remarks on the use of *well* in direct quoted speech. Since *well* can be used as what Sacks, Schegloff, and Jefferson (1974) call a *prestarter,* it is often a turn-taking signal. Its occurrence in direct reported speech serves two functions. In the first place, it is possible that the occurrence of *well* accurately reports what was uttered in the original dialogue. Secondly, as Schiffrin points out, *well* provides a shift in orientation from the speaker's own words to those attributed to the reported speaker. In the Ayr interviews there are 45 examples of *well* in direct quoted speech; of these 43 (96%) indicate a change of speaker, what Schourup calls *enquoting* (1985, 31). Sometimes this and the accompanying intonation are the only indications of such a change, as shown in the examples in (6):

(6) a. DN400   "Mrs Nicoll do you want your husband
          401   to work?"
          402   "Of course I do
          403   he's all in the ru—a nuisance at home"
          404   "*Well* that's all right"
      b. WL1880   and Davie would say to them
          1881   "*Well* lassies would you like some tatties
          1882   to take hame with you"

1883   "Oh aye"
1884   "*Well* you can fill your bags at dinner-time"

Another use that occurs in the Ayr interviews is what Schiffrin calls *embedded referent/response pairs*, where the second speaker's response depends on a response from the first speaker. Examples are given in (7):

(7) a.   (Which poems do you like best?)
      WR804   Burns' poems?
                   (Yes)
      WR805   *well* I would say "A Man Was Made to Mourn"
     b.   (What did you do first?)
      AS478   in the pits?
                   (Yes)
      AS479   *well* it was like everything else

Well also can be used in framing function, either to initiate or to close a sequence. Examples of the first kind are given in Item 8, where the utterances precede extended narratives or descriptive accounts:

(8) a. DN710   *well* I have a wee friend a Rotarian in Laguna
                        Beach
     b. WR112   *well* to gie you an idea
     c. EL3578   *well* she walked away and left him
     d. WL1395   *well* this is what happened

Examples of closing utterances are given in (9):

(9) a. NM494   *well* he discouraged me anyway
     b. WR390   *well* it's quite right that I ken
     c. EL3990   *well* he caused the only strike
        3991   that was in that stamp work
     d. WL1806   *well* it was his son
        1807   that done that

In these two uses *well* serves a bracketing function, helping to mark off narrative sections.

Within a narrative *well* is often used to highlight a critical point in the complicating action. In the first example, Laidlaw is reaching the end of a narrative about playing truant and missing an important examination:

(10) EL767   but I forgot
        768   I had to take a report card home
        769   and when I took that report card
        770   she wanted
        771   to ken

```
772    what the average mark was
773    well I couldnae tell her a lie
774    because all she would have done was
777    take me by the back of the neck
778    and run me to Newton Park School
779    so I says
780    I may as weel tell the truth and shame the devil
781    and that was a thumping
782    believe you me
```

In several places Laidlaw comments on the strictness of her upbringing, so the remark in line 773, "Well I couldnae tell her a lie," carries with it the full implications of what the consequences of telling the truth would be. This narrative came as part of an answer to a question about the kind of schooling she had received and was in support of the statements that "the standards were set far far higher than what they are today" and "we paid more attention simply because we had to." This is the only occurrence of *well* in the whole section dealing with this topic, not just the narrative; and it introduces an utterance illustrating that the "standards" had been internalized.

The second example comes in the interview with Sinclair as an illustration of a situation down the mine in which he had been really scared. The narrative begins with a long orientation section of about 70 lines and then about 15 lines of narrative proper before the part quoted in (11):

```
(11)   AS985   and in the process of doing that
       986    we had just got a hole made
       987    and I just could get my head and shoulders out
              through it
       988    that was just
       989    what like it was [demonstrates with hands]
       990    when this tremendous rumble started again
       991    and I mean it was something
       992    coming again you see
       993    well I'll tell you
       994    I jumped oot through that heid first
       995    and my neebor behind me and—
       996    the first time when I got it out through
       997    I mean it was kinna
       998    you'd
       999    to push yourself
       1000   to make your way in
       1001   well when that rumble come
```

1002   I went oot through it
1003   and I guarantee
1004   I never touched the sides [laughs]
1005   so these was wee kind of frights
1006   that you got—eh—now and again

The first *well* in line 993 introduces the metalinguistic statement that precedes the first iteration of the resolution of the action, followed immediately by the repetition of the description given first in lines 986/87 to clarify what was remarkable about "jumping oot"; and then the second *well* in line 1001 introduces the full statement of how remarkable the escape actually was.

The preceding analysis of *well* by no means exhausts all its uses; but it is sufficient to show that this discourse marker, which, as Schiffrin observes, "allows speakers and hearers to wander back and forth among levels of saying, meaning, and doing . . . searching jointly for coherence along the way" (1985, 664) is very similar in its functions in Ayr and Philadelphia, despite many other differences between the two varieties. This suggests that *well* serves a very basic function in the language—or perhaps it would be more accurate to say that it is an important member of a set of discourse markers that serve this function, since (as we shall see) several other discourse markers share certain uses with *well*.

The most similar is *oh,* with which it sometimes combines as *oh well.* In the following discussion *oh* represents a monosyllable that may be pronounced [o], [ɔ], [ɔx], or [ax], though the great majority of forms are [o].

Like *well, oh* occurs in response to questions; but the distribution between WH-questions and *yes/no* questions is very different in the lower-class interviews, as can be seen by comparing the figures in Table 10–8 with those in

Table 10.8. *Oh* in Answers to Questions

	Yes/no questions		WH-questions	
Use of *oh*	%	n	%	n
Lower-class				
With *oh*	34	62	21	22
Without *oh*	66	122	79	81
Middle-class				
With *oh*	4	7	13	10
Without *oh*	96	163	87	65
All				
With *oh*	19	69	18	32
Without *oh*	81	285	82	146

Table 10–7. For the middle-class speakers, the distribution of *oh* is very similar to that of *well;* but the number of instances is very small. As pointed out above, the overall frequency of *oh* in the middle-class interviews is only 2.7 per thousand words, compared with 7.2 per thousand words in the lower-class interviews. Moreover, only 12% of the examples of *oh* in the middle-class interviews occur in response to questions; in the lower-class interviews the proportion is 17%. The distribution of *oh* in response to questions is thus less revealing in the middle-class interviews. More significant is the pattern in the lower-class interviews where the distribution of *oh* is the reverse of that found for *well,* since *oh* occurs much more frequently in response to *yes/no* questions than to WH-questions. For the sample as a whole *oh* responses are about equally distributed between *yes/no* questions and WH-questions. This suggests that *oh* does not have the same function of maintaining coherence that was found in *well.* Schiffrin argues that *oh* is a marker of information management (1987, 73–101). Nevertheless, *oh* also has a clear function as a turn-taking signal. This can be seen, for example, in reported direct speech. There are 60 examples of *oh* in reported direct speech, all marking a change of speaker; again, this can be useful when there is no reporting verb, as in the examples in (12):

(12)   a. WR145   and the auld yin says to him
            146   "Whaur are you going?"
            147   he says
            148   "I'm going to my work"
            149   "*Oh* you got a job?"
            150   "Aye"
            151   "Whit are you daeing?"
            152   says
            153   "I'm laboring tae a bricklayer in the town"
            153   "*Oh* very good
            154   away and get your bus" he says
        b. DN2069   I said
            2070   "*Oh* many a happy night I spent in the Tower
            2071   Hotel at Elgin"
            2072   "*Oh* you wouldnae know it now mister"

It is remarkable that speakers in quoting direct speech should include items such as *oh* and *well* when these items would not be contained in any paraphrase of the propositional content of the utterance. Since it is unlikely that the reporters would remember whether the original speakers used these items, their occurrence supports Tannen's claim that quoted direct speech is *constructed dialogue* (1989, 98–133). The role of these discourse markers in quoted direct speech is to emphasize what Schiffrin calls the "participation framework" showing "the ways in which speakers and hearers can be related to their utterances" (1987, 27). On the other hand, speakers do not usually include any of the verbal discourse markers (e.g., *you know*) in quoted direct speech. The nature of quoted direct speech will be discussed in the next chapter.

Oh is also used for self-repair, but usually in the form of qualification or amplification, as in the examples in (14):

(13) a. DN710 well I have a wee friend a Rotarian in Laguna Beach
     711 *oh* about four-foot-two
  b. AM103 and then we went back to a line maybe—*oh* five or seven feet away
  c. EL1141 *oh* I went to the carpet works
     1142 *oh* I forgot aboot that wee bit
  d. EL617 and it had a pale blue sash—*oh* a ribbon aboot that width tied like—

Like *well, oh* can initiate a narrative:

(14) a. HG1390 *oh* the worst journey
     1391 I had
     1392 was
  b. WL2181 *oh* and I've been roond aboot at the racecourses and one thing and another

It can also mark the end. Laidlaw has been telling about the last day of school when she was thrown into a pond in her best clothes:

(15) EL647 so you can imagine the sight
      648 I was
      649 when I arrived home
      650 she stripped off
      651 what I had on
      652 gied me a thumping
      653 and put me to bed
      654 the sun was shining
      655 everybody else was playing
      656 and I'm lying on my bed
      657 listening to them
      658 shouting
      659 "You're het"

660   "No it's me
661   that's het"
662   *oh* I was beautiful that day

There are clearly many similarities between *oh* and *well,* but there are differences in addition to the difference in response to *yes/no* questions. Schiffrin, dealing with more interactive speech samples than the Ayr interviews, claims that "*Oh* is a marker of information management: it marks shifts in speaker orientation (objective and subjective) to information which occur as speakers and hearers manage the flow of information produced and received during discourse" (1987, 100). Because of the different character of the speech event in the Ayr interviews, it is not possible to substantiate Schiffrin's view wholly. Schiffrin does not give frequency figures for *oh* in her interviews; but if her view is correct, *oh* should be very a frequent discourse marker there, given the high frequency of *oh* in the much less interactive lower-class Ayr interviews. Schiffrin also claims that the principal role of *oh* is at the cognitive level, whereas the role of *well* is more at the interactive level. This distinction is less clear in the Ayr interviews; but since the discourse situations are different, the Ayr materials do not offer an opportunity for the careful and detailed analysis that Schiffrin presents. However, it is also possible that Schiffrin overemphasizes the differences between *oh* and *well.*

One major difference between *oh* and *well* in their distribution is that *oh* can freely combine with other elements, as shown in Table 10–9. In contrast (apart from *oh well,* of which there are 25 examples), *well* rarely combines with other items, as shown in Table 10–10.

Since *oh* freely combines with a following element and *well* rarely does and

Table 10.9. *oh* in Combination
with Other Items

Expression	Number of Instances
Oh aye	131
Oh yes	21
Oh no	18
But oh	14
Oh and	12
And oh	11
Oh but	2
So oh	1
Yes oh	1
Total	211 (= 33% of all *oh*)

Table 10.10. *well* in Combination
with Other Items

Expression	Number of Instances
And well	8
Aye well	3
Well no	3
Yes well	1
Or well	1
Total	16 (= 3% of all *well*)

since also the two occur only in the order *oh well,* never **well oh,* it seems reasonable to suggest that *oh* has a privileged position as an initial discourse marker and that *well* occupies this position only in the absence of *oh.* Moreover, since *well* is the more complex item (phonologically if not semantically), it is not surprising that it is more likely to be used when there is a need to maintain coherence. Were it not for the fact that the two can cooccur, it would have been possible to claim that the two are in a suppletive relationship, with *well* being the "strong" form of *oh.* When the two are combined, particularly in the form *och well,* the effect is often one of resigned acceptance, almost the oral equivalent of a shrug of the shoulders:

(16) a. EL3710 *och well* it didnae do ony hairm
    3711 did it
  b. WR479 *och well* I'm no offending naebody doon there
  c. WL1174 "*Oh well*" I says
    1175 "That's us
    1176 going for another game"
  d. DN1186 "*Oh well*" he said
    1187 "You're not staying at the grammar"

You Know, (You) Ken, (You) See, and I Mean

The four verbal discourse markers can also be considered together, since there are similarities (as well as differences) in their use. The lower-class speakers use these discourse markers with a frequency of 13.1 per thousand words, almost three times as much as the middle-class speakers (4.5 per thousand words). The only one of these discourse markers that is used with almost equal frequency is *you know* (in the lower-class group 2.3 per thousand words, 2.1 per thousand in the middle-class interviews). The lower-class speakers, how-

ever, have an alternative in *(you) ken*—since *ken* as a main verb has the same
meaning as *know,* as illustrated in (17):

(17) a. HG1427   and I didnae *ken* much aboot the underground
      b. WL446   "Bobby you *ken*
         447   where I'm going" says I
      c. EL3064   "Hoo do you *ken*
         3065   whit wire's which?"
      d. WR758   "You *ken*
         759   where John Wayne's buried and aw?"

The examples in (17) illustrate the use of *ken* as a main verb and consequently
are not cases of *(you) ken* as a discourse marker.

  You know and *(you) ken,* however, do not function in exactly the same way
as discourse markers, so it is necessary to consider them separately. *You know*
can occur in initial position in the clause (18a and b), in medial position (18c
and d), or in final position in the clause (18e and f). By *initial position* is
meant either the first position or immediately following a coordinating conjunction or a discourse marker such as *well;* by *medial position* is meant any
position preceded and followed by any constituent other than a coordinating
conjunction or another discourse marker; and by *final position* is meant a
position followed by no constituent other than a terminal tag such as *and that.*

(18) a. AM1439   *you know* it wasn't any worse
      b. EL812   *you know* you'll maybe not believe me
      c. NM1360   flour would be in the other one *you know* for the
                   baking
      d. WL77   and I could see *you know* the hunted look on his
                   face
      e. JM571   it was a wonderful feeling *you know*
      f. AS194   that were played later on *you know* and that

The distribution of *you know* is given in Table 10–11. In both groups (but
especially in the lower-class interviews) there is a preference for final posi-

Table 10.11. Distribution of
You Know (%)

Class	Initial	Medial	Final
Lower	16	19	64
Middle	25	34	41
All	20	25	55

Table 10.12. Distribution of *(You) Ken* (%)

Class/Respondent	Initial	Medial	Final
Lower-class	6	13	81
Rae	4	10	87
LC w/o Rae	17	24	60

tion. (Huspek found a similar distribution [1989, 666].) As shown in Table 10–12, the pattern for *(you) ken* is similar except for the effect of Rae's interview, which accounts for 78% of the examples of *(you) ken*. Unlike *you know*, *you ken* can be reduced to *ken*. (Obviously, *know* would be homophonous with *no*, which could lead to confusion.) In the lower-class interviews as a whole, *you ken* is reduced to *ken* 82% of the time; but this figure is misleading. Rae uses *ken* 92% of the time, compared with 47% in the other lower-class interviews. It is, therefore, reasonable to note that for most of the sample *you ken* is distributed like *you know;* but the fact that Rae's use is rather different shows that it would be unwise to claim that there is no difference between the two.

Östman suggests that the prototypical meaning of *you know* is that "The speaker strives towards getting the addressee to cooperate and/or to accept the propositional content of his utterance as mutual background knowledge" (1981, 17). Östman argues that the emphasis is on the striving and that "taken to its extreme, this says that a speaker will use *you know* when the addressee does NOT know what the speaker talks about" (p. 17). An illustration of this can be seen in the interview with Sinclair, where there is a clear difference between his use of *as you know* (which is not a discourse marker) and *you know* (which is).

(19) a. AS883   *as you know*
      884   just prior to the war they started
      885   doing away with the tramcars up and down the
             country
    b. AS2310   well *as you know*
      2311   things was getting bad just the first year of the war

Since Sinclair and I are about the same age and had been in the same part of Scotland at the time, he could reasonably assume that these statements referred to mutual background knowledge; and he introduces them with the formulaic phrase *as you know*. By way of contrast, in the following example he knows that I have never been in the Maxwell colliery and could have no firsthand experience of it:

 (20) AS1246 I mean when in the Maxwell colliery for instance
    1247 you can stand at the foot of the road and look up
    1248 *you know* it's just like
    1249 looking up inside a big chimney
    1250 I mean you can only see
    1251 as far as the light would show you

Sinclair knows that I do *not* know what it is like in the colliery, and his elaboration in line 1250 introduced by *I mean* shows that he is fully aware of this. In another example, Sinclair is describing one of the kinds of night out that the miners would have at a hotel with the excuse of a presentation to one of the older miners. There would be drinking and a singsong:

 (21) AS1928 and then half-time the landlady or that came in with
       maybe a big ashet of skinny tatties again and butter
       and that
    1929 *you know* you enjoyed these things
    1930 and it made you
    1931 drink mair in the second half

This is a good example of what Östman refers to as "a striving towards attaining a Camaraderie between the speaker and the addressee" (1981, 19). Sinclair is saying, in effect, that if I had been present, I would have enjoyed it too.

 Sometimes initial *you know* is equivalent to the rhetorical question "Would you believe?"—as in this example from Laidlaw's interview:

 (22) EL3803 but oh vandalism noo's—
    3804 it's ridiculous
    3805 there nothing really
    3806 you could dae with them
    3807 toilets at Whitletts Cross a thing
    3808 that never—we never had
    3809 I mean we fought hard
    3810 the ratepayers fought hard
    3811 to get them
    3812 *you know* they're a shambles
    3813 you can't go into them

Laidlaw does not assume that I know that the toilets are a shambles; on the contrary, she expects me to be surprised.

 Medial *you know* is sometimes used as self-repair to add a further elaboration or qualification (cf. Schourup 1985, 120–25):

(23) NM1624 seeing her with her shawl *you know* wrapped in her shawl

      DN659 he was an athlete *you know* the PT instructor you see

      AS2221 but it still was a wee village

      2222 the old hooses and everything like that *you know* the old hall and that

As Schourup (1985, 122) points out, *you know* is used as a repair involving a radical change of meaning.

Östman claims that *you know* can occur within constituents, for example, *I'm closing the—you know—window* (1981, 168). No examples of this kind occurred in the Ayr interviews. *You know* always occurred at constituent breaks. Huspek has a similar observation (1989, 667).

The most frequent use of *you know* is in clause-final position. Östman observes that "utterance final *you know* can have either a Declarative or an Interrogative contour" (1981, 21). If by *Interrogative contour* Östman means a high rising nucleus, there are very few examples in the Ayr interviews, all in the middle-class interviews. More common is a sustained or slightly rising nucleus, but most common is a low falling one. The sharply rising nucleus asks for agreement, as in this example from MacDougall's interview:

(24) AM1200 but I think

      1201 there is this feeling

      1202 that well we can't do it for ourselves

      1203 we've got

      1204 to get the experts in London

      1205 to do it *you know*

By his intonation MacDougall is asking me to agree with the basic proposition that lack of self-confidence is a problem in Scottish politics. The more common sustained or falling intonation is more like a request for sympathy or understanding:

(25) a. IM454 I was finding it difficult *you know*

      b. DN2623 there no tenpenny ones now *you know*

      c. EL185 maybe through my mother's mistake I was rather tightly—held to a tight schedule *you know*

      d. MR533 my family was young *you know*

Schiffrin claims that *you know* is a marker of metaknowledge about what speaker and hearer share and also about what is generally known (1987, 268). She emphasizes its role as an information state marker rather than its interactional function; but since she includes many examples of what I consider main verb uses of *know,* her analysis is based on different assumptions.

In its full form *you ken* functions much like *you know,* but in its reduced form *ken* it can occur within constituents in a way that *you know* does not:

(26)   a. WL813   some of them would've *ken* ta'en on the coal
         b. WR451   and they just turned it on hard
              452   and *ken* brushed aw these things
         c. WR443   oh they washed aw the dirt oot for *ken* Hogmanay
         d. WR548   maybe even playing at rounders or—*ken* tennis

The high frequency with which *(you) ken* is found in Rae's interview makes it rewarding to look in more detail at its use there. In terms of the larger organization of the interview *(you) ken* is fairly evenly distributed: 24% of the tokens occur in the first quarter of the interview, 22% in the second, 30% in the third, and 24% in the fourth. The tokens, however, are not equally distributed throughout the text. In narrative passages *(you) ken* is less likely to occur. There is one narrative of 22 lines with no examples of *(you) ken* and another of 26 lines with only one example. Since *(you) ken* occurs on average every five to six lines in Rae's interview, at least eight examples would have been expected in the two narratives. When *(you) ken* does occur in narratives, it often accompanies orientation or background clauses, as in (27):

(27)   WR303   I mind o yin instance
         304   actually an auld man
         305   it was always a steak pie
         306   that was the kinda feed for these types of things *ken*
         307   steak pie and tatties *ken*
         308   so it was to be seven-and-six
         309   in auld money *ken*
         310   before they changed and that
         311   but here with one or two no turning up
         312   they had
         313   to increase the cost to ten shillings
         314   "Oh tae hell" this auld yin says
         315   "Get the manager"
         316   "Whit's wrang sir?"
         317   "Well tae stert with" he says
         318   "the pie's too dear
         319   and there nae links in it
         320   and we've decided
         321   to eat it under protest" [laughs]

The clauses in lines 306, 307, and 309 are background information not essential to the narrative structure; and it is characteristic of Rae's narrative style to use *ken* for such information rather than in narrative clauses.

Table 10.13. Distribution of
(*You*) *see* (%)

Class	Initial	Medial	Final
Lower	30	5	66
Middle	13	9	78
All	26	6	69

Another pattern that emerges is that my questions often follow an utterance by Rae that ends in *ken*. Approximately a third of my questions follow immediately on *ken* and this is most apparent in the middle part of the interview. It is as if I took his use of *ken* at these points as a sign that he was relinquishing the floor. Since this happens twice as often as would be expected, it is clearly a factor in the dynamics of the interview. Like *you know*, *(you) ken* functions as a marker of interactional solidarity.

(You) see shows a similarity in distribution to *you know* and *(you) ken*, with the majority of occurrences in final position, though *(you) see* is less likely to occur in medial position. The distribution is shown in Table 10–13. Like *you ken*, *you see* can be reduced, in this case to *see*. Unlike the reduced form *ken*, however, the reduced form *see* occurs primarily in initial position, as can be seen in Table 10–14. For the full form *you see*, the predominance of final position can be clearly seen in Table 10–15.

Table 10.14. Distribution of *see* (%)

Class	Initial	Medial	Final
Lower	84	0	16
Middle	100	0	0
All	85	0	15

(You) see differs from *you know* and *(you) ken* in that the emphasis is less on the assumption of a shared perspective than on the need for the hearer to grasp the point. In initial position the sense is "I want you to see (understand)," and in final position it is "I hope you have seen (understood)." As a result, it is a more distancing discourse marker, with the emphasis on the speaker's telling and the listener's paying attention. It is an *I–you*–relationship rather than a *we*-relationship. When the speaker says *you know* or *you ken*, the implication is one of a shared perspective: "we know," "we ken." When the speaker says *you see*, the emphasis is rather on the listener's role in understanding what is being said. The examples in Item 28 show the kind of explanatory clauses that are often introduced by *(you) see:*

Table 10.15. Distribution of
You see (%)

Class	Initial	Medial	Final
Lower	22	3	75
Middle	10	9	81
All	19	5	76

(28) a. WL464   *you see* between your shots you're supposed
            465   to have a certain time
      b. AS927   *you see* between each seam of coal you have a
                     layer of rock
      c. EL2029 well *you see* that kind of thing annoys me
      d. DN666   *see* he was the most alert man in the—in the
                     crowd

In an extended description it is not uncommon to have a sequence of *you see*s
in fairly close succession, as in this extract from Sinclair's interview:

(29)   AS1274 I mean there was no such thing as
          1275 having
          1276 to shovel your coal
          1277 because your coal used to rumble right down to the
                     bottom *you see* and that
          1278 but after they got up to their distance
          1279 wherever they were going to *you see*
          1280 or the maximum where they could go tae
          1281 they could drive along *you see*
          1282 and these roads
          1283 that they drove along up to a hundred feet
          1284 well you come back to the first one
          1285 you had *you see*
          1286 and you used to go right along that hundred feet and
                     take all the coal out up that hundred feet *you see*

It is perhaps not surprising that Sinclair, Lang, and Nicoll should be responsible for the majority of the uses of *(you) see* in the interviews, since they are
the most devoted explainers; 88% of the examples of *(you) see* occur in these
three interviews. Lang has also the highest proportion of narratives, but *(you)*
see is less frequent in narratives; 7 (27%) of the 26 narratives in Lang's
interview have no examples of *(you) see*. When *(you) see* is used in a narrative, it is often in an orientation clause, as in the example in (30):

(30) WL323 and I remember
      324 there were an ice-cream man
      325 come to the pit on a Friday
      326 and you got a double nougat
      327 you know these wafers
      328 thruppence
      329 and you got a cigarette with it
      330 a cigarette—and thruppence
      331 three old pennies for a double nougat
      332 and this was a great thing on a Friday night at the pit
           you know
      333 and I put the cigarette in my top pocket
      334 and I went doon the pit with it on the Monday
           morning
      335 so what happened
      336 I seen this cigarette
      337 and I took it
      338 and rubbed it oot like that
      339 *you see* it was an offence
      340 to go doon with cigarettes in your pocket
      341 although I had nae matches
      342 but if it was found
      343 oh you had cigarettes
      344 you would have been fined for that
      345 oh aye you would have been fined for that
      346 and yet in that pit doon there you can take cigarettes
           with you
      347 some pits you can smoke *you see*
      348 and others you don't

In lines 339 and 347 Lang is providing me with the information that I need in order to understand the point of the narrative. In the Ayr interviews *(you) see* functions as the kind of information state marker that Schiffrin (1987) found *you know* to be in Philadelphia.

I mean is the first-person discourse marker corresponding to *(you) see*. Where *(you) see* places the emphasis on the listener's task of comprehending, *I mean* stresses the role of the speaker in making the message clear. *I mean* essentially occurs only in initial position, but its reference is generally to the immediately preceding discourse. On the surface, at least, *I mean* is an example of what Schiffrin (1980) calls "meta-talk," that is, talk about talk, since it suggests an attempt to clarify what has been said; but only occasionally is the clause that follows *I mean* in any sense a paraphrase of what has gone before, as it is in (31):

(31)   DN1149   and I had no—uh—preferment at all
           1150   *I mean* I wasn't academically clever

In line 1150 Nicoll rephrases the proposition that his euphemistic statement in line 1149 had largely obscured. Most of the uses of *I mean* are not of this kind. More often, *I mean* is used to introduce an explanation or elaboration:

(32) a. MR186   no there nobody
           187   got pocket money in these days
           188   because nobody'd money
           189   to give them pocket money—nobody
           190   *I mean* wages wasnae big in these days
    b. WL1600   oh I worked with the milk
         1601   fae I was aboot five-year-auld with John Aird fae Broadheid
         1602   well he was at Kirkhall first
         1603   and he shifted to Broadheid
         1604   and eh—*I mean* it was always a help sort of thing
         1605   we got two or three coppers
         1606   for running with the milk in the morning

Sometimes the elaboration simply takes the form of repetition for emphasis:

(33) a. EL706   if I made a mistake
           707   or if I was reported for anything
           708   I got a thumping
           709   and *I mean* I got a thumping
           710   it wasn't just a clip on the ear and sent ootside
    b. EL1375   I think
         1376   they're broader
         1377   it's not even a dialect
         1378   it's slang
         1379   *I mean* it's slang
         1380   they're using
         1381   and mixed with that's obscene language

Sometimes *I mean* is used to sum up a point:

(34) a. WL1237   Mr Patterson he was a gentleman
         1238   he was a giant
          1239   he was a giant of a man

          1240   (No I've just heard his name) did you know him?
          1241   he was a gem
          1242   *I mean* his life went on public work
          1243   it did that
    b. AS3611   I remember a man
          3612   that used to go in
          3613   swimming
          3614   and he always had his bunnet on
          3615   he even had his pipe
          3616   going and that
          3617   *I mean* these were characters and that

Sometimes, but not very frequently, *I mean* is used for self-repair:

(35) a. AS3223   and even in my younger day—*I mean* before the
                   war
      b. WR817   he could talk Latin and French
           818   but it was—
           819   *I mean* it wasnae—
           820   the schooling he got was spasmodic

The use of *I mean* varies greatly, with Sinclair being responsible for more than
half the examples in the whole sample. The following extract illustrates his
repeated use of *I mean* in an eloquent passage in which he is striving to give a
vivid impression of the experience of going down the mine for the first time.

(36) AS626   I remember the first morning
         627   going in the cage
         628   and going down the pit
         629   *I mean* the sensation
         630   of going down
         631   *I mean* you go away
         632   and the cage goes away
         633   and *I mean* you just feel your breath
         634   going away fae you
         635   *I mean* it's no like a lift
         636   it goes doon at some tune
         637   and the sensation was
         638   after you were maybe about halfway doon
         639   you felt
         640   as if you were going back up
         641   that was just the kind of situation
         642   it was *you see*

The four verbal discourse markers, *you know, (you) ken, (you) see,* and *I mean,* make an important contribution to the dialogue that the interviews contain. *You know* and *(you) ken* serve mainly to create an atmosphere of shared perspectives and attitudes. The *you* does not have a strong deictic function referring to the listener alone but rather aims to bring the listener into an inclusive relationship with the speaker in a shared *we*-perspective. *(You) see* focuses much more clearly on the listener's role as a counterpart of Grice's (1975) cooperative principle for the speaker; the listener's cooperative responsibility is to pay attention and try to understand what is being said. *I mean* draws attention to the speaker's efforts to make things as easy as possible for the listener. While all four are strongly involved in sustaining an effective interaction, *(you) see* and *I mean* are more narrowly associated with information states.

Anyway and Mind You

Anyway is included by Quirk and his colleagues among "the concessive conjuncts" that "signal the unexpected, surprising nature of that is being said" (1972, 674). The sense is often "contrary to what you might think or have expected":

(37)	a. NM494	well he discouraged me *anyway*
	b. IM281	it was very hard
	282	to refuse
	283	and—eh—*anyway* we accepted it
	c. HG1601	and there nothing
	1602	that you could buy really much *anyway*
	d. AS1726	I took it *anyway*

Mind you has very similar force:

(38)	a. DN278	*mind you* it'll be ordinary
	b. WL1024	it's a pity *mind you*
	c. MR174	I believe
	175	we had more freedom *mind you*
	176	than what they have nowadays
	d. HG1639	it was a hard life
	1740	but *mind you* it didnae dae you ony—
	1641	I don't think
	1642	it did a body ony herm

While *anyway* is used by both social class groups, *mind you* is more common in the lower-class interviews (though it is not frequent even there). *Anyway*

occurs both clause-initially and clause-finally; but in the lower-class interviews it is more likely to occur in clause-final position, while in the middle-
class interviews nearly 80% of the examples occur in clause-initial position.

Of Course and In Fact

Quirk and his colleagues list *of course* and *in fact* as "attitudinal disjuncts that
assert the truth of their sentence" (1972, 674). *Of course* is like an opaque
alternative to *you see* with much the same function except that it is perhaps
slightly more polemical with the sense of "surely you understand that." It
occurs in initial, medial, and final position, but primarily in initial position. In
the lower-class interviews, 94% of the examples of *of course* are in initial
position, compared with just under 70% of the middle-class examples. Only
the middle-class interviews have examples of *of course* in medial position.
Illustrative examples are shown in (39):

(39) a. WL1147   *of coorse* it was miners against miners sort of
                   thing
      b. IM070   *of course* I went to school for a few years
      c. JM846   and he *of course* was first choice
      d. AM599   this was *of course* one of the drawbacks
      e. WR675   that was a nonexistent baun *of course* ken
      f. JM2349   it goes back to the primary school *of course*

In fact occurs predominantly in initial position in both sets of interviews,
with less than 15% of the examples in final position and only two examples in
medial position. The sense of *in fact* is an elaboration or qualification of an
earlier point, as illustrated in (40):

(40) a. WL2566   oh you got a ferret for half a croon at that time
            2567   *in fact* you got them
            2568   for taking away
      b. AS1950   as I said before
            1951   I mean it was an experience
            1952   to go down in the cage in the first time
            1953   *in fact* not only the first time
      c. AM153   that was the first time
            154   I'd been out of Scotland *in fact*
      d. JM1177   but never a question of in-with-the-boot
            1178   and two or three watching
            1179   *in fact* two or three helping

Now and Here

Schiffrin (1987) suggests that *now,* as a discourse marker, transfers its temporal deixis to *looking forward.* In other words, *now* draws attention to something that is to follow. Quirk and his colleagues call *now* in this function "a transitional conjunct" that "leads to a new stage in the sequence of thought" (1972, 667). Obviously, in such a function *now* can occur only in initial position—unlike temporal *now,* which can occur at a variety of points in the clause. Clear examples of forward-looking *now* are rare in the interviews; one occurs in the interview with Sinclair:

(41)   a. AS1594   and then you go back to the bottom of the list then
        1595   and you take your turn
        1596   *now* as you know
        1597   in every type of work you get drones
        1598   even in a beehive there are the drones and everything like that

More often *now* is looking *back* to a *previous* statement:

(42)   a. EL1788   her father's in—doing a life sentence for murder
        1789   he murdered her mother last year
        1790   *now* that's deprived home
        b. HG1692   when the war broke oot
        1693   and you started
        1694   to earn money
        1695   they come roon and telt us
        1696   that the government had stopped them
        1697   fae—fae daeing this
        1698   but you could put in at least four shillings in
        1699   *now* this is a cock-and-bull story

Both forward-looking and backward-looking uses of *now* are illustrated in this extended example from Sinclair's interview:

(43)   AS2861   *now* such is the incident
        2862   that happened this year at Wembley
        2863   when the one supporter run on the field and all that
        2864   and he started
        2865   scrapping and that
        2866   *now* that wasn't a Scotchman
        2867   that was an Englishman
                (I didn't know that)

2868   that's true
2869   that was an Englishman
2870   *now* that's never been said much about in the papers at all
2871   but why?
2872   *now* it's never been said much about in the papers at all in any way
2873   but if it had been a real Scotchman
2874   I can assure you
2875   it'd never been out of the papers
2876   but that was an Englishman that fellow
2877   *now* that is fact

Sequences of a particular discourse marker such as Sinclair's use of *now* in Item 43 are not uncommon (see Östman 1981, 41), and Sinclair often seems to fall into a pattern where he uses one discourse marker several times in quick succession. It is also part of Sinclair's generally repetitive style (see Chapter 12). *Now* in line 2861 points forward to the argument that is to follow but the other instances of *now*, particularly that in line 2877, refer back to what has been said earlier.

Here, like *now*, transfers its deixis, in this case from locative to temporal. The sense is "at this point," but there is also a suggestion of surprise at what happened:

(44)   a. HG347   and it was two or three days
           348   till I was retiring
           349   and the wife she went into the Buroo aboot it
           350   and *here* they'd my wrang age doon
   b. DN1252   now the men were paying five pound to my father for a suit
         1253   and *here* this comes in
         1254   and you get two suits for the five pound

With the exception of the two in Nicoll's interview, all the examples occur in the lower-class interviews. (Although transcribed as *here*, it might be possible to make a case that it is really *hear*, in the sense "listen to this.")

Terminal Tags and Hedges

Dines (1980) examines the problems in accounting for the use of terminal tags such as *and things like that*, *and all this*, and *and everything*. Dines found such tags to be more frequent in the speech of working-class Australians than among middle-class speakers. The same is true of the Ayr interviews. The tags to be considered are shown in (45).

(45) *Terminal tags*
      and that
      and that there
      and all that
      and everything like that
      and one thing and another
      et cetera
      or that
      or that there
      or all that
      or something like that

There are 361 examples of these tags in the interviews, and 344 (95%) occur in the lower-class interviews; but 293 of these (85%) occur in the interview with Sinclair. The high frequency of these tags in Sinclair's interview is clearly an idiosyncratic feature of his speech; he uses approximately 14 tags per thousand words compared with just over 1 per thousand in the rest of the lower-class interviews. (See Macaulay 1985 for a discussion of Sinclair's use of terminal tags.) Even without Sinclair's figures, however, terminal tags are more frequent in the lower-class interviews than in the middle-class interviews, so that Sinclair is taking to an extreme a feature of lower-class speech. What is surprising about this is that in certain other respects (e.g., his use of relative clauses) Sinclair deviates from the other lower-class speakers in the direction of middle-class norms; in this case he moves in the other direction by exploiting so vigorously a feature that is not characteristic of middle-class speech. It is perhaps significant, however, that he makes frequent use of the tag *et cetera,* which is not used by any other lower-class speaker but accounts for a third of the tags in the middle-class interviews.

Dines argues that in her sample of Australian English such terminal tags have a set-marking function "to cue the listener to interpret the preceding element as an illustrative example of some more general case" (1980, 22). Some examples can be interpreted this way, as shown in (46):

(46) a. EL1891   these bingos and social clubs *and one thing and another*
      b. WR833   maybe dae a tour o the Burns country Dumfries *and that*
      c. AS1153   and you were taking wee samples *and all that*
      d. AS496   you get singles trebles doubles dross *et cetera*

The great majority of tags, however, are much harder to interpret as in any
way set-marking:

(47)   a. HG908   if I had her cleaned up *and that*
       b. WR292   aye but it—it was a cheap holiday *and that*
       c. MR088   I didn't feel it tight *and that*
       d. AS035   when he died *and that*

Sinclair also, surprisingly, often makes use of more than one tag, which
further weakens the set-marking effect:

(48)   a. AS3171   they get on their nerves *and one thing and another*
                   and that
       b. AS2143   as far as women *et cetera and that* was concerned

Sinclair's use of terminal tags suggests that they function pragmatically more
like "hedges" (G. Lakoff 1972). (49) shows some phrases that occur in the
Ayr interviews as hedges:

(49)   *Hedges*
       sort of
       sort of thing
       sort of style
       kind of (kinna)
       kind of thing(s)
       kind of style

There are 226 instances of these phrases in the interviews, and they occur
slightly more frequently in the middle-class interviews (2.1 per thousand
words) than in the lower-class interviews (1.7 per thousand words). Particularly characteristic of middle-class speech is the use of *sort of*, which can be
used within constituents, as can be seen in (50):

(50)   a. WG1467   but—they're *sort of* showing the people to the
                   people
       b. AM1560   I can *sort of* read here and just get some peace
       c. AM1107   because I tended
          1108   to *sort of* come back home
       d. IM277   I *sort of* don't know

The lower-class interviews account for only about 20% of the examples of *sort
of;* but both social class groups use *kind of* with approximately the same
frequency, and it also occurs within constituents:

(51)   a. HG930   but what I think would be the *kinna* faut o the thing
                   here

    b. WR545   it was *kinna* hamemade fun in them days ken
    c. NM503   well I *kind of* took a—a simple course
    d. IM662   I *kind of* can look forward to the time

It is interesting that Sinclair, whose use of terminal tags is so prodigal, uses relatively few of these hedge phrases. There are other signs that the use of these features is idiosyncratic: two speakers (Nicoll and Lang) are responsible for 83% of the examples of *sort of thing* while two others (Muir and Laidlaw) provide 77% of the examples of *kind of thing(s)*. There does not appear to be any difference in meaning or function between *sort* and *kind* in these phrases.

Coordinating Conjunctions

As mentioned above, Schiffrin includes the coordinating conjunctions *and, but,* and *so* among the discourse markers. These forms were not included in the discussion of the other discourse markers because their syntactic role constrains their discourse function. Table 10–16 shows the distribution of these forms for the two social class groups.

Table 10.16. Number and Frequency of Coordinating Conjunctions

Respondent	and		but		so	
	n	/1,000 wds.	n	/1,000 wds.	n	/1,000 wds.
Lower-class						
Gemmill	357	38.8	116	12.6	44	4.8
Sinclair	550	26.2	163	7.8	25	1.2
Lang	412	25.8	102	6.4	75	4.7
Laidlaw	299	23.9	94	7.5	60	4.8
Ritchie	92	21.9	47	11.2	3	.7
Rae	71	13.7	32	6.2	18	3.5
All LC	1,781	26.2	554	8.1	225	3.3
Middle-class						
Nicoll	519	34.1	137	9.0	43	2.8
Menzies	147	31.3	39	8.3	6	1.3
MacGregor	229	29.7	61	7.9	13	1.7
Muir	106	24.7	44	10.2	10	2.3
MacDougall	223	21.0	71	6.7	10	.9
Gibson	174	20.7	77	9.2	50	6.0
All MC	1,398	27.5	429	8.4	132	2.6

The figures in Table 10–16 show that there is no essential difference between the two social class groups in the frequency with which they use these
conjunctions; nor are there obvious explanations for the individual differences
in the frequency of use. As Schiffrin (1987) observes, the variety of functions
and can serve make its greater frequency to be expected. In particular, *and* can
occur in successive clauses in narrative or descriptive accounts, which is rare
with *but* and never happens with *so*. Repeated use of *and* is found in both
lower-class and middle-class interviews. Examples are given in (52) and (53):

(52)	WL2067	we went on the Queen Elizabeth
	2068	we left Southampton
	2069	*and* we went over to le Havre
	2070	*and* we went from le Havre to Ireland
	2071	*and* we left Ireland
	2071	*and* then across the Atlantic tae America tae New York
	2072	*and* we were conducted tours roond New York sort of thing
	2073	we were all roond New York
	2074	*and* then we flew back
	2075	come back across in the plane
	2076	*and* we got to Gatwick
	2077	I think
	2078	it was Gatwick aye Gatwick
	2079	*and* I think
	2080	she thought
	2081	I was running away fae her at that time
	2082	I don't like the aeroplanes no no no no
	2083	I was ready for the toilet
	2084	when we landed

The whole narrative from line 2069 is carried by clauses introduced by *and;*
the clauses in lines 2074, 2076, and 2078–79 are asides or elaborations. *And*
does not occur in the last three lines, where Lang explains his behavior.

(53)	NM1442	before we had—eh—stainless steel knives
	1443	for using
	1444	the knives
	1445	we had
	1446	were steel—the cutlery knives
	1447	*and* we had a round—
	1448	*and* we turned the handle

1449   *and* you put the knives in
1450   *and* there was an emery friction thing
1451   *and* it cleaned them beautifully
1452   it was one of the great things
1453   to be allowed
1454   to put the knives in
1455   once you grew bigger
1456   you held them in
1457   *and* you birled this thing round
1458   *and* they came out—shining

Again, although this is a "reminiscence" (in Horvath's [1985] terms), lines 1452–56 are by way of an aside to the main account given by the clauses introduced by *and*.

It is rarely that *but* introduces successive clauses, since the function of *but* is to mark a contrast with the preceding statement. In the following example from Nicoll's interview the difference between the successive clauses introduced by *but* is a matter of scope:

(54)   DN1947   I left school at fourteen
       1948   I was never very well educated
       1949   I regret that now
       1950   *but* it was my choice
       1951   my father and mother wanted me
       1952   to stay on
       1953   *but* I made my choice
       1954   *but* I love
       1955   to sit in company
       1956   and listen to good English well spoken

This is a good example of Nicoll's rhetorical style with the parallelism of the first three clauses echoed by the repetition in lines 1950 and 1953. The first *but* in line 1950 refers back to the statement in line 1949; the second in line 1953 refers to the statement in lines 1951/52; and the third in line 1954 refers to the whole notion contained in lines 1947/53, with the sense that despite the fact that it was his own choice to leave school at fourteen and not become well educated, he still loves to hear English well spoken. It is, of course, possible to avoid a repetition of *but* by using *however,* as in the following example (also from Nicoll's interview):

(55)   DN2221   I knew it all you see
       2222   *but* I was wrong you see
       2223   *however* the point to the—the highlight to the story
                was—

In line 2222 Nicoll is concluding an episode in what is to be an extended narrative, and in line 2223 he is making an attempt to get back to the main story line again. The contrast between *but* and *however* marks this break more clearly than would have been the case if *but* had been used in line 2223.

So is never used in successive clauses; but it can recur at critical points in a narrative, as illustrated in this example from Lang's interview:

(56)	WL657	my neighbor come
	658	and he says to me
	659	"Willie you'll need to come up to Curly"
		[5 lines omitted]
	665	he says
	666	"Cannae see him
	667	I think he's buried"
	668	*so* away I goes
	669	running
	670	and this big stane was doon you know
	671	I says
	672	"Where is he?"
	673	"Oh" he says
	674	"he's in there"
	675	I says
	676	"Oh naw"
	677	"Aye he's in there"
	678	*so* up over the top of this
		[17 lines omitted]
	697	he was lying in here
	698	*so* I says to him
	699	"Are you aw right?"
	700	"Aye" he says
	701	"I'm no bad"
	702	says I
	703	"Well I don't ken hoo"
	704	says I
	705	"Put your haund roond my neck
	706	and I'll pu you up"
	707	in this narrow space
	708	*so* I shouted
	709	to get a streetcher and splints you see
		[11 lines omitted]
	722	"One two three"
	723	and we lifted

724   *so* we lifted the streetcher
725   and he was left
726   lying on the pavement
727   it was rotten
728   you see the streetcher was rotten
       [6 lines omitted]
735   we just lifted the streetcher
736   and he was left
737   lying there
738   *so* we had to make do with jaickets

The first *so* clause in line 668 marks Lang's response to the news that Curly is buried; the second in line 678 gives his more specific action in response to more details about the location; and the third in line 698 again refers to Lang's next task, which is to determine whether Curly is fit to be moved. Having ascertained that he is, Lang then proceeds in line 708, also introduced by *so*, to lift him up. Lang then describes the preparations for carrying Curly on a stretcher; line 724, introduced by *so*, refers to this attempt; and line 738 marks the resolution of the problem. Each of the clauses with *so* refers to a significant act by Lang, either alone or with help from others. This use of *so* as a marker of subdivisions in a narrative can be found in both lower-class and middle-class interviews.

I have looked at the use of discourse markers. Since such items are frequent in casual interactional encounters, it is interesting that they should be so widespread and varied in a dyadic situation where one of the participants was exceptionally reticent and eager to yield the floor. This suggests, on the one hand, that such features are such a fundamental part of everyday language that they occur freely even in the unusual setting of a dyadic, tape-recorded interview with a stranger; on the other hand, their frequency can be seen as a confirmation that the unusualness of the interview situation was not a severely constraining factor.

I have shown that there are social class differences in the use of discourse markers but also that there are other differences related to genre. Moreover, the use of discourse markers seems to be sensitive to individual style, with some speakers showing highly idiosyncratic use of certain features. The connection between these features and other aspects of individual style will be considered in Chapter 12. It will be seen in the next chapter how speakers display an awareness of certain discourse markers when reporting the speech of other people.

I I

The Use of
Quoted Direct Speech

I shall consider here how speakers in relating events often report what someone said, supposedly repeating the actual words said. This is traditionally
known as *oratio recta* (e.g., "He said 'That won't affect me' ") as opposed to
reported speech, *oratio obliqua* (e.g., "He said that would not affect him"). I
shall refer to *oratio recta* as *quoted direct speech*. It is one of the ways in
which "as animators, we can convey words that are not our own" (Goffman
1981, 150).

It is clear that the use of quoted direct speech is not necessary to convey the
propositional content of the utterance, as can be illustrated by an example
from the interview with one of the lower-class speakers, Andrew Sinclair:

(1) AS368   and—eh—when I was waiting on the milk
     369   eh—the farmer came oot
     370   and he says
     371   "That you left the school noo Andrew?"
     372   says I
     373   "It is"
     374   he says
     375   "You'll be looking for a job"
     376   says I
     377   "Aye"
     378   he says
     379   "How would you like
     380   to stert here?"
     381   well I would have liked

382   to have started at the farm
383   I did like farm work

The examples of quoted direct speech are the sequences shown in Item 1 in quotation marks. Sequences of this kind in natural conversation virtually never occur with the words *quote* and *unquote* or finger gestures (as occasionally seen in lectures) (Goffman 1981, 161). Speakers indicate that they are representing another speaker in a number of ways: by a reporting verb such as *he says;* by prosodic features, particularly intonation; and sometimes by attempting to mimic a form of speech that is different from their own. In (1), for example, the sequence "That you left the school noo Andrew?" in line 371 is bracketed by *and he says* in line 370 and *says I* in line 372; and it is also uttered with an interrogative intonation contour that is separate from what precedes or follows. (There is no attempt to represent the farmer's speech as in any way idiosyncratic; rather, it has the characteristics of a local speaker.) All the examples of quoted direct speech in (1) are similarly bracketed except for the last in lines 379–80, where the change of speaker in line 381 is indicated by the turn-taking marker *well* (cf. Schiffrin 1987). There is thus no doubt at any point in Item 1 as to the identity of the speaker or the referent of the deictic *you*.

The information conveyed in Item 1 could have been expressed as in (2):

(2)   1   when I was waiting on the milk
       2   the farmer came oot
       3   and he asked me
       4   if I'd left the school
       5   I said
       6   I had
       7   and he said
       8   I'd be looking for a job
       9   and when I said yes
       10   he asked
       11   if I would like
       12   to start there on the farm

or even more briefly:

(3)   1   when I was waiting on the milk
       2   the farmer came oot
       3   and asked me
       4   if I'd like
       5   to work there
       6   now that I'd left school

Extended examples of reported speech such as (2) are rare in ordinary spoken discourse, and it is not clear why anyone would want to give the information in that type of explicit form. The summarized version in (3) is a better candidate for an alternative; and examples like this do occur in the interviews, generally with the middle-class speakers, as in this example from John MacGregor's interview:

(4)   JM1115   and we had a long chat. . . .
        1118   and—eh—he asked me
        1119   to come to South Africa
        1120   and he said
        1121   I could either come into the army with him or go with
               the K____in—eh—physical training

In this example the use of deictic pronouns after the reporting verb *he said* shows that this is clearly reported speech (*oratio obliqua*), with the narrator referred to by *I* and the reported speaker by *him* (though the contrastive use of *come* [with him] and *go* [with another organization] retains the orientation of the original speaker). Reported speech is much less common than quoted direct speech in the interviews; and since the latter is generally more verbose, it is legitimate to ask what purpose is served by this extra effort on the part of both speaker and listener. It is reported speech, however, that has received more attention in linguistics (see, e.g., the articles in Coulmas 1986).

Voloshinov observes:

Earlier investigators of the forms of reported speech committed the fundamental error of virtually divorcing the reported speech from the reporting context. That explains why their treatment of these forms is so static (a characterization applicable to the whole field of syntactic study in general). Meanwhile, the true object of inquiry ought to be precisely the dynamic interrelationship of these two factors, the speech being reported (the other person's speech) and the speech doing the reporting (the author's speech). After all, the two actually do exist, function, and take shape only in their interrelationship, and not on their own, the one apart from the other. The reported speech and the reporting context are but the terms of a dynamic interrelationship. This dynamism reflects the dynamism of social interorientation in verbal ideological communication between people (within, of course, the vital and steadfast tendencies of that communication). (1986, 119)

Voloshinov is thinking of literary works, as his reference to "the author" indicates; but his remarks are equally pertinent to the study of naturally occurring oral narratives.

Functions of Quoted Direct Speech

Indirection

In the narrative from which (1) comes, Sinclair is giving a long answer to the question "What did you do when you left school?" The narrative is surprisingly eloquent and well-organized (see Macaulay 1985 for a fuller discussion of the whole narrative) and leads on to an account of Sinclair's first jobs in the coal mine. As Sinclair is to make clear at many points later in the interview, he was by his own account respected as a good worker. In Item 1 he shows how the farmer asked him if he would like to work there before Sinclair had even said he was looking for a job. This suggests that the farmer recognized Sinclair's potential and wanted to get in first with the offer of a job. The narrative episode ends with the information that Sinclair had left the job on the farm not because of any shortcomings on his part but because of a dispute between his parents and the farmer over compensation for an injury he had received. The dialogue quoted in (1) shows Sinclair, who was barely 14 years of age, being treated with respect by the farmer almost as if he were an equal. This is not conveyed so effectively by the paraphrases in (2) and (3). One of the functions of quoted direct speech is to be able to convey information implicitly that it might be more awkward to express explicitly. Or else, as Goffman observes, the speaker, by repeating the words said by someone else, "means to stand in a relation of reduced responsibility for what he is saying" (1974, 512). In Item 1 Sinclair is also following what Pomerantz has called the constraint on avoidance of self-praise (1978, 88–92). Indirection as a mode of speech is just beginning to be recognized (cf. Brenneis 1986; Brown and Levinson 1987; Duranti 1984). The narrative from which (1) is taken occurs fairly near the beginning of the interview. It is a remarkably long response to a question that could have been answered by a single sentence, e.g., "I worked on a farm for a bit." Instead, Sinclair, in a narrative that extends to 84 lines, provides a great deal of information about himself and his family's circumstances, partly as background information in the orientation sections. The quoted direct speech is not an integral part of the narrative in the sense that the farmer's precise words are essential to the point ostensibly being made (cf. some of the examples below). The farmer's attitude, however, conveyed through the quoted direct speech, is an effective preparation for Sinclair's later account of himself as a miner, working and living "among the cream of human beings in the country" and his despair when he was temporarily out of work.

Embedded Evaluation

An important use of quoted direct speech is to provide what Labov (1972, 372–73) has called embedded or internal evaluation. Labov argues that the evaluative portion of any narrative explains why the story is reportable, or worth telling. He uses the term *external evaluation* when the narrator gives this justification explicitly as a comment in his own role as narrator and the term *internal evaluation* when the point of the story is conveyed through the story itself, including the remarks of another person. There are many examples of internal evaluation in the narratives told in the interviews. One occurs in Ella Laidlaw's account of her first job in a carpet factory at the age of 16. Her mother, who worked in the factory, had got her a job in what she termed "one of the nicest parts"—the finishing room. Laidlaw soon found that the other workers would swear at her if she did anything wrong and were not sympathetic to the problems of a learner. She put up with it for about two weeks and then went to tell her mother she was quitting:

(5)	EL1234	and I went up to my mother
	1235	my mother worked in the dyehoose then
	1236	and I says
	1237	on the drying machine[1]
	1238	and I says to her
	1239	"I'm no going back in there"
	1240	she says
	1241	"How are you no going back in?"
	1242	I says
	1243	"I'm no going back in
	1244	You should hear the swears of them in there"
	1245	she says
	1246	"They don't swear in the finishing room"
	1247	I says
	1248	"Do they no?
	1249	You want to come and hear them?"
	1250	so there were a big wuman worked with her
	1251	she was an Italian girl—Ina Solotti
	1252	and Ina Solotti says
	1253	"Who said the feenishing room didnae swear Mary?"

[1]Despite the reporting verb *I says,* it is clear from the intonation that line 1237, "on the drying machine," is not quoted direct speech. Instead (as line 1238 shows), it is a false start, because Laidlaw wants to add an important detail about where her mother works. See Tannen 1989, 135–66 on the significance of details in conversation.

1254   my mother says
1255   "Me"
1256   "Well" she says
1257   "we've nothing on them
1258   away you go and listen to them at their door" she says
1259   "and you'll get the shock of your life
1260   Who was swearing at you hen?" she says
1261   and she was a great big wuman
1262   she was aboot eighteen stone
1263   I says
1264   "Jean MacPherson and Dorothy Thompson"
1265   "Ah well" she says
1266   "I'll go oot to them"
1267   so on goes the cap again
1268   and away she crosses the yaird
1269   and what she gied these two lassies
1270   was naebody's business

This passage illustrates the effective use of repetition as a rhetorical device in lines 1239, 1241, and 1243 in which *no going back in there* is repeated with appropriate deictic shifts by both speakers (cf. Tannen 1987, 1989). (There are even more striking examples elsewhere in the narrative, but they are not relevant to the present discussion.) Laidlaw at many places in the interview stresses that she was brought up very strictly by her mother. In this narrative she shows that her mother was not always right in her assessment of situations, and she brings out the full force of this judgement by putting it in the mouth of Ina Solotti. Laidlaw has already described the conditions under which she had to work, but the exchange between Ina Solotti and Laidlaw's mother emphasizes the extent of the latter's misjudgement: "We've nothing on them." Note Ina Solotti's question in line 1253: "Who said the feenishing room didnae swear, Mary?" This echoes the mother's remark in line 1246, "They don't swear in the finishing room." This is the kind of clarification request preceding disagreement that Pomerantz notes in conversational exchange (1984, 71). She also remarks: "Disagreement components may also be delayed within turns. Conversants start the turns in which they will disagree in some systematic ways. One way consists of prefacing the disagreement with 'uhs,' 'well's,' and the like, thus displaying reluctance or discomfort," (p. 72). Ina Solotti prefaces her disagreement in lines 1256–57 with *well*: "'Well,' she says, 'we've nothing on them'." It is significant that speakers in reporting the speech of others include discourse markers such as *well*. (See below on other examples of the use of discourse markers in quoted direct speech.) The function of embedded evaluation is one of the most frequent uses of quoted direct speech, presumably because responsibility for the remark is clearly attributed to someone other than the speaker. As Voloshinov observes:

> Reported speech is regarded by the speaker as an utterance belonging to *someone else*, an utterance that was originally totally independent, complete in its construction, and lying outside the given context. Now, it is from this independent existence that reported speech is transposed into an authorial context while retaining its own referential content and at least the rudiments of its own linguistic integrity, its original constructional independence. (1986, 116, emphasis original)

Voloshinov is referring to *oratio obliqua*. In *oratio recta* the quoted utterance may retain more than the rudiments of its original construction. The next section deals with situations in which attention is drawn to specific details of the quoted direct speech.

Mimicry

While all examples of *oratio recta* have some features of imitation in them, there are cases where the speaker is clearly not just reporting what someone said but also clearly imitating the original speaker as closely as possible (cf. Polanyi 1982, 163).

Wallace Gibson is a theatrical producer and an amateur actor, and much of his use of quoted direct speech is apparently to show off his skill as an actor in directly imitating others. In relating how his teachers at school were surprised at his success as Malvolio in the school production of *Twelfth Night,* he reports them as saying, "Gibson we didn't know you had it in you"; and his intonation and tone of voice catch the Scottish teacher's patronizing and sarcastic manner very effectively. This is polyphonic speech (Bakhtin 1973) at its richest. Gibson is both conveying the attitude of the teachers to the boy and the boy's (and adult's) reaction to them. There is no need for explicit comment on either. (It should be remembered, of course, that he was speaking to someone who had experienced such teachers, so that there no need to spell it out; on the contrary, it was horrifyingly effective.)

Mimicry can also be used more explicitly in dealing with attitudes toward language. In (6) Gibson describes a visit to the house of the parents of a member of the amateur theatrical company:

(6) WG1075   his parents—are—ah—are—em—sort of
      1076   [coughs] at least his father is a self-made man
      1077   who's now quite wealthy
      1078   his mother is a very—uh—brilliant char—mind and

1079   but the father—still—speaks—sort of—
1080   I was up at the house recently in Troon
1081   and—eh—we were wanting
1082   to use his—recording gear for something
1083   and he was
1084   [imitating] *"Oh just a minute noo*
1085   *I can get the plug"* and so on
1086   *"Now here we are*
1087   *now boys"*
1088   quite a wealthy man
1089   but he speaks like that
1090   but mother [imitating] *talks like that*
1091   *"You know Mr Gibson"*
1092   and Douglas said
1093   *"Oh it irritates me*
1094   *she's a nice person mother*
1095   *but she puts on this voice"*
1096   he says
1097   *"I like my father better*
1098   *because he doesn't cover up"*

In this passage Gibson imitates three speakers, the father, the mother, and Douglas, the son—quite successfully, though the transcript (the imitated sections are in italics) does not bring this out. Gibson's use of the three forms is similar to the use of parodied sociolects found in Scottish comedians (Macaulay 1987a). It is clear that Gibson identifies with the form of speech he uses for Douglas, which lies between the affected form used by the mother and the lower-class form used by the father. Gibson goes on to tell of a visit to a miner's family; and he imitates their form of speech and also, briefly, that of an English curate's wife whose speech is in high contrast to that of the miner's family. Gibson himself claims to use a wide variety of speaking styles from "an almost RP at times . . . down to speaking quite Scots."[1] His use of quoted direct speech, which is quite frequent in his interview, allows him to demonstrate this in a number of quite different situations. His mimicry of lower-class speech is quite effective, employing intonation and voice quality as well phonologically marked forms. His mimicry of the mother's affected speech and of the Englishwoman is very brief and closer to parody than a realistic imitation.

[1]Received Pronunciation (RP) is a term used to refer to the prestige accent in England. See Macaulay (1988d) for a discussion of the influence this term has had on sociolinguistic investigation.

Quoted direct speech allows the speaker to show knowledge of linguistic features that he or she does not normally use in the interview situation. Duncan Nicoll is the only middle-class speaker who occasionally uses the negative clitic *–nae* instead of the expected *–n't*. As was shown in Chapter 5, all the lower-class speakers variably use forms such as *wasnae, doesnae,* and *couldnae* where the middle-class speakers use *wasn't, doesn't,* and *couldn't.* Most of Nicoll's examples come in quoted direct speech, as shown in (7):

(7)   DN119   I said
             120   "Would you like
             121   to go to the castle?"
             122   "Oh but mister" they said
             123   the castle's shut"
             124   "Oh" I said
             125   "I know the man—
             126   that's got the key"
             127   "Oh that would be fine
             128   but we *cannae* walk"
             129   it's a fair walk up to the castle
             130   I said
             131   "Oh but I—
             132   we'll get the cars up to the front door"
             133   "But you *cannae* get
             134   doing that mister"
             135   I said
             136   "Oh I know the man
             137   that lets you up to the door" you see
             138   so we got them up
             139   and got the door opened

Here in lines 128 and 133 Nicoll demonstrates his familiarity with lower-class speech. Presumably all the middle-class respondents could have done this, but it is only Nicoll and Gibson who do. In this case Nicoll appears to be having a little bit of fun at the expense of "the old decrepit people" (as he had earlier described them) who have come to visit the castle. He knows that I know that he is the man with the key and with the authority to let the cars go up to the door; but he tells the story as if he had concealed this information from the pensioners. It could be argued that he is simply being modest with them in not explaining to them that he is the man in charge, but the way in which he quotes their remarks suggests that he is really scoring points off them because of their ignorance. The pensioners' statements underline their helplessness in the face of physical handicap ("we cannae walk") and unresponsive authority ("the castle's shut," "you cannae get doing that"). Their addressing him *mister* indicates politeness but no great respect. If they had known he was the man in charge, they would not have addressed him in that way. The story that Nicoll is telling is actually one that shows him in a good light. He was kind to these old people and they appreciated it; but the quoted direct speech, both his own and that of the old people, reveal an undercurrent of a slightly patronizing attitude. This is brought out by the conclusion to the narrative:

(8)   DN149   we came to the going away
          150   and I'm standing with—er—
          151   one of my friends—a very dear friend of mine had on
                   her arm a terribly crippled bent old woman
          152   and as she passed me
          153   she said
          154   "Good night Your Highness"
          155   [nods slightly] just her head
          156   she couldnae move—just her head
          157   it was wonderful
          158   it made the day

This is perhaps how Nicoll wanted to be addressed: *Your Highness,* not just *mister.* Although he tells it as a joke, it may help to explain the undercurrent of the earlier use of quoted speech.

Among the middle-class speakers, Gibson and Nicoll display a particular ability to mimic; and their narratives are populated by a wide variety of individuals speaking in character, where the purpose of the mimicry is to draw attention to the form of speech. As it happens, in the Ayr interviews the middle-class speakers mimic the lower-class speakers, never the reverse. This asymmetry may be accidental, since I have recorded examples of lower-class speakers mimicking middle-class speech elsewhere; but it is also probable that middle-class respondents feel more comfortable imitating lower-class speech in the presence of a middle-class interviewer than lower-class speakers in the complementary situation. It is also likely that the salient features of lower-class speech are easier for middle-class speakers to reproduce than the less marked characteristics of middle-class speech are for lower-class speakers. Goffman (1974, 537) suggests that there are "rules of mimicry" that vary "from culture to culture and within a speech community from one category of speaker to another." This is a question that deserves more attention than it can be given here.[1]

---

[1]Mitchell-Kernan describes the form of mimicry known as *marking* in Afro–American speech: "The marker attempts to report not only what was said but the way it was said, in order to offer

Taboo Expressions

Quoted direct speech can also include the use of taboo expressions that might not otherwise be considered appropriate in the interview situation. On several occasions one of the lower-class speakers, Sinclair, an ex–coal miner, hints at stories but claims, "Oh there's some of them too crude to be told." (It is at such moments that one wishes the interview were taking place in the evening over a glass or two of beer rather than in the afternoon with a cup of tea.) Almost at the end of the interview, however, Sinclair refers to the kind of "great characters" there used to be in the neighborhood:

    (9)   AS3632   you were maybe sitting
            3633   and you say
            3634   if you done this or that
            3635   and Davie used to say
            3636   "Ah but if the wee bird didnae shite
            3637   it would burst"
            3638   you know I mean this was one of his great sayings

There is only one other taboo expression in Sinclair's long interview, and it also comes in quoted direct speech. Sinclair would not have used the word *shite* (the long vowel is normal in Ayrshire speech) as part of his nonnarrative speech in the interview—at least not without some kind of apology. As Goffman asks in connection with what he calls "censorship rules," "In quoting another's use of curse words and other taboo utterances some license is provided beyond what the speaker can employ on his own behalf, but where does this license stop?" (1974, 539). The few examples of taboo expressions that occur in the Ayr interviews come in quoted direct speech, but they are all relatively mild. The above example from Sinclair is the strongest.

Authenticity

The implication of the use of quoted direct speech in narratives is that the speaker is reporting what was actually said, since otherwise it would be a

---

implicit comment on the speaker's background, personality, or intent. . . . His performance may be more in the nature of parody and caricature than true imitation. But the features selected to overplay are those which are associated with membership in some class. . . . Perhaps more than anything, marking exhibits a finely tuned linguistic awareness in some areas and a good deal of verbal virtuosity in being able to reproduce aspects of speech which are useful in this kind of metaphorical communication" (1972, 176–79). She also observes, "In a linguistic community which is bilingual or bidialectal, the code in which messages are conveyed is likely to be highly salient both to members of the community and to the ethnographer" (p. 163). Although I would not call the situation in Ayr bidialectal, certain aspects of lower-class speech are highly salient; and it is these features that are likely to be used in mimicry.

flagrant violation of the Gricean maxim of quality. The quoted speech in Item 9 is not part of a narrative but an example of one of Davie's "great sayings." In examples such as Item 9 it is reasonable to suppose that the reported words are close to (or identical with) the original, since the saying itself is the point of the example. In narratives, however, the quoted remarks are part of the action of the narrative and have same standing as the other narrative events reported. The validation for Nicoll's report that the old lady said, "Good night Your Highness" is exactly the same as that for the indication that she nodded her head. Nicoll is reporting on what he heard and saw. Since the remark is short and the point depends upon the form of address, the listener has no reason to question its validity.[1] With longer utterances, the validation may be less obvious; and in some cases it may be clear that what is quoted could not have been uttered (Tannen 1989). The main interest, however, is not the historical accuracy of the reporting, which is unverifiable, but the plausibility of the quoted speech. The problematic cases are those in which it is not easy to explain the form of the quoted speech, since it must sound convincing if it is to have the right effect, whether or not the speaker is constructing the dialogue.

In certain situations it is reasonable to suppose that the words themselves were so memorable as to stick in the speaker's memory. There is some evidence in experiments testing recall of dialogue that people have a substantial memory for surface structure and not just meaning (Bates, Masling, and Kintsch 1978; Hjelmquist 1984). If this can happen in experimental situations where there is no emotional involvement, it is perhaps not surprising that speakers should be able to recall specific remarks that made an impression on them.

Tannen (1986, 1989) argues against this view. She claims that "what is commonly referred to as reported speech or direct quotation in conversation is constructed dialogue, just as surely as is the dialogue created by fiction writers and playwrights" (1986, 311). She gives several examples of quoted direct speech that could not have been uttered, for example, thoughts that the speaker reports as not having been uttered and quoted remarks that are paraphrases or summaries of what was said. This raises the question of the definition of quoted direct speech. Tannen includes everything that the transcriber chooses

---

[1]Sometimes the speaker makes it clear that the speech is not being quoted exactly. For example, Nicoll introduces a story by saying (DN2691/93): "And uh this morning I was in a chatty mood and I said, 'Oh hello Mr So-and-so'." Obviously, this is not what Nicoll said; Nicoll is not using the man's actual name, presumably either because he has forgotten it or because he does not think it appropriate to use it. Even if he has forgotten the name, it does not follow that his recollection of the rest of the conversation is inaccurate, since with advancing age names become increasingly difficult to recall.

to put in quotation marks, even when the speaker does not claim to be reporting verbatim. One of her examples is: "and this man is essentially saying 'We shouldn't be here because Imelda Marcos owns this restaurant' " (1989, 113). Here the speaker explicitly denies that the actual words are being reported. It would be better to restrict the notion of quoted direct speech to cases where the speaker seems to be vouching for the accuracy of the actual words used, but even then there are doubtful cases. Usually the attributed remarks are introduced by some form of the verb *say*.[1] Unfortunately, even in these cases, it is as hard to show that quoted direct speech is always constructed as it is to show that it has been recalled accurately. There are some situations in which the reported speech is more plausible and some in which it is less. In (10) Willie Lang is telling of an accident that happened in the pit to a man who had been having problems because his wife was not paying their bills:

(10)   WL758   I'll never forget the look on his face
       759   when I went up
       760   he was lying you see
       761   so I row'd up his troosers [i.e., rolled]
       762   "Oh" I says
       763   "Jackie, your leg's broken"
       764   "Oh no, Willie
       765   it cannae be
       766   it cannae be broken"
       767   says I
       768   "Jackie your leg's broken
       769   make no mistake aboot that"
       770   "Oh" he says
       771   "it cannae be" he says
       772   "this is the end of the quarter"
       773   this was the Cooperative
       774   "The end of the quarter as sure as I never move"
       775   that's the words
       776   that wee man said
       777   and I could see you know the hunted look on his face
       778   it just couldnae be
       779   because he'd

[1]There are no examples of *like*, as in "And we're like 'What are you talking about?' " (Tannen 1989, 116). It would be interesting to explore whether there is any difference between such remarks introduced by *like* and those prefaced by some form of *say* in those varieties where *like* is used in this way.

780   to pay this Cooperative there at the quarter
781   that was a fact

In addition to Lang's explicit claim, "That's the words that wee man said," the point of the story is the problem with paying the bills, not the injury itself; so some of the quoted speech is integral to the story and not merely a way of making it more dramatic. The essential point is not whether Lang is recalling the actual words Jackie used (although it is possible that he is; cf. Goodwin 1982, 801). The rhetorical structure of the narrative with its counterpoint of repetitions clearly make this a well-told story:

(11)   763   Jackie your leg's broken
        766   it cannae be broken
        768   Jackie your leg's broken
        765   it cannae be
        766   it cannae be broken
        771   it cannae be
        778   it just couldnae be
        772   this is the end of the quarter
        774   the end of the quarter as sure as I never move
        780   to pay this Cooperative there at the quarter
        775   that's the words
        776   that wee man said
        781   that was a fact
        758   I'll never forget the look on his face
        777   and I could see you know the hunted look on his face

Whether he is remembering or constructing the quoted speech, Lang has succeeded in telling the story effectively; and the fact that quoted speech is fully integrated into the telling gives it a certain plausibility.

In some cases the question of plausibility is more complicated. Laidlaw narrates the story of how her father died unexpectedly: "He just lay doon on the settee and turned over and that was him gone." She goes on to tell how her mother had left her fathering sitting in the garden while she went into town to shop for groceries:

(12)   EL3587   and it was an exceptionally good afternoon
        3588   and she put him out in a basket chair
        3589   sitting at the window ootside in the garden
        3590   she went in on the one bus
        3591   and came back on the same bus
        3592   because the conductress says to her

3593   "Thought you said
3594   you were going for messages" [shopping]
3595   she says
3596   "So I was"
3597   "Well" she says
3598   I'm awful glad
3599   I'm no waiting on you" she says
3600   "You couldnae have got much
3601   because you've got the same bus back"
3602   "Ach well" she says
3603   I don't like the idea
3604   of leaving him too long"
3605   and she went up the road
3606   she noticed
3607   his basket chair was there
3608   but he wasnae there
3609   she never thought anything aboot it
3610   because it was too warm
3611   she thought
3612   he'd naturally gone inside
3613   and when she went in
3614   he was lying on the settee
3615   and she's auld-fashioned very tidy very smart
3616   everything had to go in its place
3617   she took off her coat
3618   hung it up
3619   put away her shopping bag
3620   and she says
3621   "It's rather early for wer tea—wer dinner
3622   so I'll go and ask him
3623   if he wants a coffee"
3624   and she made the coffee
3625   and she went through
3626   and shook him
3627   to ask him
3628   if he wanted tea
3629   and he dropped off the settee in front of her
3630   and she just—
3631   her mind just broke
3632   and she's never known
3633   what it is since

This is an extraordinary story. Presumably, Laidlaw herself was not present during any of the events she narrates. The only source could have been her mother. So even if "her mind just broke," she must have been coherent enough to tell what had happened if we are accept the quoted direct speech at its face value. It is also a story full of detail:

(13)   3617   she took off her coat
        3618   hung it up
        3619   put away her shopping bag

Laidlaw also relates her mother's thoughts, both indirectly, "She noticed his basket chair was there," "She never thought anything aboot it," "She thought he'd naturally gone inside" (with the transposed tense *he'd gone*) and in *oratio recta*, "It's rather early for wer tea" and so on. The latter is a good candidate for "constructed dialogue." Laidlaw's mother presumably addressed these remarks (or thoughts) to herself, since there can have been nobody else present. It is highly unlikely that the repair from "tea" to "dinner" is the mother's; it was probably Laidlaw's for my benefit, since she would know that I would have my dinner in the evening rather than in the middle of the day, and she might have thought that I would think of tea as "afternoon tea."

It is the section in lines 3591–3603, however, that is most intriguing. Laidlaw could only have known about this exchange if her mother (or the conductress) had told her. The story could have been told without these lines, since they do not advance the action, which has already been fully described: "She went in on the one bus and came back on the same bus." What the dialogue with the conductress provides, however, is the dramatic anticipation of the tragedy, "I don't like the idea of leaving him too long." Once again, the quoted speech has an important role to play in the narrative, though it is structurally less crucial to the story than in the previous example from Lang. By the use of quoted direct speech Laidlaw has transformed what would otherwise have been a straightforward third-person account of her mother's actions on the day that her father died into a dramatic narrative in which the perspective varies with different speakers. It is likely that much or all of this dialogue has been "constructed" in Tannen's terms; but even so, it is an impressive demonstration of her narrative skills.

Translation

The character of the quoted speech may come from the narrator rather than from the speaker to whom the speech is attributed. Gemmill had earlier explained that when he was young he had had no vacations, though after

World War II things had been different. Here he is describing his first visit to London with his daughter:

(14)   HG1301   eventually we come to a corner
         1302   and here there's addresses in this
         1303   so we're staunning [i.e., standing]
         1304   looking at this
         1305   when this wuman came along and said
         1306   what were we looking for
         1307   and we're looking for somewhaur
         1308   to stay the night
         1309   "Where do you come fae?"
         1310   "Scotland"
         1311   "You're no feart of coming here without
         1312   somewhere to stay"
         1313   so she gi'en us half a dozen addresses

It is highly unlikely that the woman whose speech is quoted is herself Scots. If she had been, the exchange would probably have included some reference to that or questions about which part of Scotland Gemmill and his daughter were from. If the woman had been Scots, that would have been a reportable fact, since it would have been a coincidence the first person who spoke to them was herself Scots. It is even more unlikely that anyone not from Scotland would say "Where do you come fae?" ("Where do you come from?") and "You're no feart" ("You aren't afraid"). In this case it seems obvious that Gemmill is reporting (or constructing) informal speech in the form that is normal for him and not for the speaker to whom the remarks are attributed. This is the mirror image of mimicry. For Gemmill to present the speech of the woman in an appropriate form for a Londoner he would have had to abandon his normal form of speech. Although he presumably could have attempted this, he does not do so. The reason is probably that unlike the cases of mimicry, the woman's form of speech is not integral to the story. On the contrary, it is her friendly behavior that is important. Attempting to imitate her form of speech might have implied a critical attitude to that kind of speech or appeared to mock her. Instead, Gemmill employs the most neutral form of speech—his own—regardless of the realism of the quoted speech. Although they are not bracketed by markers other than intonation, there is no way to interpret lines 1309–12 as other than quoted direct speech. There is an obvious contrast with the reported speech in lines 1305–6, with the transposed pronoun *we* compared with the deictic *you* in line 1311. This seems to be a clear example where the use of quoted direct speech cannot be taken as evidence of the form of the original utterance, though the content of the quoted remarks is plausible enough. There is a second example a little later in the interview describing another visit to London when Gemmill reports the ticket seller at Euston Station as saying:

(15)   HG1458   there mair folk nor you [i.e. more people than you]
       1459   got to go back
       1460   to start their work

This would be more plausible if reported of Glasgow Central Station, as it certainly is not a typical specimen of London speech. Again, however, the phonological and syntactic form of the speech are irrelevant; it is the content of the utterance that is important for the narrative. By using his normal form of speech Gemmill avoids distracting attention from the main point. It is perhaps worth adding that I do not recall finding these examples of quoted direct speech at all remarkable when Gemmill was telling me the stories. It was only on analyzing them that the discrepancy between the remarks and the speaker to whom they were attributed struck me. In this sense, Gemmill's tactic worked. On the other hand, I was quite aware at the time of the significance of the mimicry employed by Gibson and Nicoll.

Self-Quotation

Finally, it is interesting that speakers do not use quoted direct speech only to convey the views of others. Speakers quite frequently quote themselves, even when there is no other dialogue and the information could easily have been conveyed in a different form. The example in Item 16 comes from the interview with Nicoll where he is explaining his problems as a church elder visiting the families in his district, which he admitted was "not the best end." Nicoll has found that it is difficult to get enough quiet to be able to "give them the story of the church," as he put it:

(16)   DN1734   and uh it's a job
       1735   when you go in
       1736   to get them
       1737   to put off the television
       1738   the minister finds this quite often
       1739   that uh he's fighting against the television
       1740   but if he does—
       1741   I can get—
       1742   and I'm sure
       1743   he does it too
       1744   "Now listen

1745   I'm only in for ten minutes
1746   will you just switch off that box
1747   I know
1748   it's your favorite program
1749   but switch it off
1750   I'll go away in ten minutes
1751   and you can get it again"
1752   and you get them
1753   and you give them the story of the church

This is probably an illustration of the kind of speech Nicoll would give, rather than a specific example, though it is also possible that he might have used these exact words. He certainly uttered them in a tone that sounded plausible, though it would not necessarily have been effective. On many occasions in the interview Nicoll makes it clear that he is an effective speaker and that he is aware of this ability and how important it has been to him in various aspects of his career. Later he explains how he can talk to people on a wide range of subjects:

(17)   DN2020   and you've got
        2021   to be able
        2022   to talk bookie
        2023   you've got
        2024   to be able
        2025   to talk pub
        2026   to talk everything
        2027   but having had the width in the hotel business the
                C____business
        2028   then you get used to
        2029   talking to people
        2030   and finding a way

In (17) the first two uses of *you* in lines 2020 and 2023 are equivalent to the indefinite *one*, but in line 2028 Nicoll is referring to himself in the way that the middle-class speakers sometimes refer to themselves by *one* as a less personal form of *I* (see Chapter 6).

In (16) Nicoll is simply illustrating his skill in a difficult situation in much the same way as he might quote someone else that he admired. This kind of self-quotation thus becomes a kind of distancing whereby Nicoll can present himself as an actor in a scene even though the others present are not given the opportunity to speak (cf. Goffman 1974, 519–20). Nicoll is the speaker who has the highest proportion of quoted direct speech in the total sample; and more than half of that consists of his own remarks, though usually his remarks form part of a dialogue with other speakers.

Problematic Cases

Usually, the examples of quoted direct speech are so convincing that it is hard to believe that the speakers are "constructing" them. There are, however, occasions when the use of quoted direct speech is less convincing. Later in the story about his job on the farm than the part given in Item 1 Sinclair reports his response to the farmer's question:

(18)   AS407   and—eh—I said
       408   I would like it fine anyway
       409   and well he says
       410   "Think
       411   and ask your mother" you see—eh
       412   "See
       413   what they say at it
       414   and tell her" he says
       415   "you've fifteen shillings a week"
       416   so I went hame with the milk
       417   and I told my mother and faither
       418   what was said and that
       419   and they said
       420   "Well if you want
       421   to start there
       422   you can start"
       423   so I sterted work on the ferm the next day—eh—and
                that

Although, as pointed out earlier, this was an important event in Sinclair's life, the form of the farmer's remarks is not a crucial part of the story, and nothing critical hinges upon their being reported accurately. Yet they sound convincing both in the preposition in *see what they say at it* and in the pronoun switching in *tell her*. In Sinclair's interview and in other lower-class interviews it is clear that it is the mother who is responsible for the family finances, and she is the one who gets any money earned by the children; so the pronoun switch in line 414 is appropriate. Both mother and father are to be consulted about the possibility of Sinclair's working on the farm, but it is the mother who would need to know how much money he would earn. If Sinclair is not repeating what the farmer actually said, he has managed to create a plausible imitation.

The remark attributed to his mother and father, "Well if you want to start there you can start" is less convincing. In the first place it is an improbable remark in itself. It seems more like the summary of an exchange which might have included the direct question, "Do you want to start there?" Its implausibility is reinforced by the introducing verb *they said,* since it is highly unlikely that "they" both uttered it (in unison? in sequence?). There is also the contrast between *start* in lines 421–22 and *sterted* in line 423. This is the reverse of a sequence earlier in the same story:

(19) AS379 "How would you like
     380 to *stert* here?"
     381 well I would have liked
     382 to have *started* at the farm

As pointed out in Chapter 4, the form *stert* for *start* is the traditional dialect form; and Sinclair fluctuates in his use of such forms both here and elsewhere in the interview. The question that arises in lines 421–23 in Item 18 is whether the contrast between *start* and *stert* is deliberate. In (19) it is natural that the farmer should be reported as using *stert* in speaking to the boy, while Sinclair, in addressing his middle-class interviewer, uses *start;* but it is less obvious why the use should be reversed in lines 421–23 in Item 18. One possibility is that Sinclair's parents (or at least his mother) actually used the form *start* and not *stert* and that he is accurately reporting what she said (or might have said, since the truth is not the issue). Another possibility is that the use of *start* in lines 421–22 indicates either in the original discourse or in the narrative a formality of diction appropriate for a serious decision. The third possibility— and the most likely—is that Sinclair is summarizing the gist of what they said in a form that he feels is appropriate in addressing a middle-class listener. It is also possible that there is no narrative significance to the switch and that the two forms are in this respect in free variation for Sinclair. This last possibility is the least satisfying explanation, but there is a similar alternation in another story that Sinclair told near the end of the interview. (The whole story is reproduced elsewhere (Macaulay 1985, 117–19.) In this story Sinclair three times refers to a remark made by his father about not buying him a drink:

(20) AS3512 my father bought a round of drinks after the meal
     3513 there wasnae one for me you see
     3514 and one of the men happened
     3515 to comment
     3516 he says
     3517 "Bob" he says
     3518 "you forgot the boy"
     3519 he says

3520   "No" he says
3521   "I didnae forget the boy" he says
3522   he says
3523   "If he wants a drink" he says
3524   "he'll buy it
3525   but I'm no buying him *one*"

Since Sinclair was an adult at this time and a member of the group that was on their way to a race meeting, it would be normal for someone buying a round of drinks to include him. One of the men obviously notices this and "happens" to comment on it. (The references to *the boy*, meaning "your son"/"my son," are normal forms of familial reference for the lower-class speakers; see Chapter 6.) This is a story that Sinclair told me to illustrate how things had changed since he was young when "you knew your place" and did not expect to be treated as an equal by adults immediately after leaving school. The point of the story is that his father would not object if Sinclair had a drink but that he was not going to encourage him by buying him one. Sinclair was apparently anxious that I should get the point of the story because he repeats the main point twice:

(21)   AS3538   and he says
        3539   "Well I'm no buying him a drink
        3540   because I don't believe in it
        3541   but if he wants *yin*"
        3542   as much as to say
        3543   "Well if he wants *yin*
        3544   he can have *yin*
        3545   but I'm no buying him it"
                [22 lines omitted]
        3567   he didnae say
        3568   "Don't drink"
        3569   but he says
        3570   "I'm no buying you *one*"

Lines 3523–25 in (20) give the remark as first quoted in the context of the story, and the indefinite pronoun is *one*. Lines 3539–45 in (21), although reported as quoted direct speech, are in fact a paraphrase of what his father said (and meant); the indefinite pronoun, used three times, is *yin*. Lines 3568–70 in (21) are a rephrasing of the paraphrase with a change of the personal pronoun from *him* to *you*, and the indefinite pronoun is again *one*. This is a puzzling situation that it would have been easier to explain if *yin* had been used in (20), (the original version) and *one* for the other examples in (21), where Sinclair is clearly rephrasing the original remark for my benefit. As it is, the switch in (20–21), like that in (20), raises questions about the significance of the differences. Since Sinclair is the lower-class speaker who makes the greatest accommodation to the interview situation, it is tempting to look for significance in the kind of switches shown here; but the explanation is not obvious.

A different kind of example comes from the interview with Laidlaw, who is the other lower-class speaker to show extensive variability during the interview. She quotes her singing teacher at primary school:

(22)   EL517   when you went for singing
          518   he used to say
          519   "Annie Laurie?
          520   Annie Laurie would dee
          521   she would die in her box" he says
          522   "If she heard youse
          523   there nane of youse can sing" he would shout at us

Laidlaw uses *nane* (for *none*) only twice elsewhere in the interview, and this is her only use of *dee* (for *die*). It is interesting that she should quote a teacher as using these forms and also the stigmatized plural form *youse*. It is minimally credible that a primary teacher at that time could have used these forms in addressing the children, but it would not have been the expected form of speech. It is also possible that Laidlaw has translated what the teacher said into how she might have heard it as a child even if the teacher had used middle-class forms. Later in the same story, however, Laidlaw tells how she had got into trouble for making up words about this teacher and being overheard singing them by the teacher himself:

(23)   EL564   the following morning we goes in
          565   he's standing
          566   and he says
          567   "Right Miss Dunlop [Laidlaw's maiden name]
          568   upstairs to Mr Walker"
          569   that was the headmaster
          570   up the stair we went
          571   and that boy—unfortunately that boy Patterson was among them
          572   and he
          573   "Please sir we didnae sing it
          574   it was just her"
          575   they were that loyal [laughs]

576   because we were all petrified of him
577   "Youse were all singing"
578   "No sir we didnae say the words
579   it was Ella Dunlop
580   that said the words"
581   so Ella Dunlop got into trouble
582   and when Mr Walker the master he says to me
583   "Repeat what you said Ella"
584   and when I repeated it
585   he couldnae strap me
586   for laughing
587   he says
588   "You must admit" he says
589   "Mr Kerr" he says
590   "there's something
591   she must be a bit of a poet" [laughs]
592   I wasnae a bit of a poet
593   I thought I was being funny

Once again, Laidlaw reports the teacher as using *youse* (in line 577); but she reports the headmaster as using a very different kind of language. Although Laidlaw does not try to mimic him, she uses a different voice quality, suggesting a more typically middle-class speaker. She also reports the headmaster as using epistemic *must*, a form that is rare in the lower-class interviews and more common in the interviews with the middle-class speakers. This suggests that the contrast between the teacher's language and the headmaster's is not accidental and therefore that the use of *nane, youse,* and *dee* in (22) may be significant.

There remains a puzzle in (22). The change from *dee* in line 520 to *die* in line 521 is hard to interpret. It does not appear to be Laidlaw's repair, changing to a middle-class form of speech more appropriate for a teacher, since she quotes him as using *nane* and *youse* in the next two lines. It is unlikely to be the teacher's own self-repair, since line 521 is not an obvious repair of line 520. In fact, "She would die in her box" seems more like a misstatement for something like "She would turn in her grave." If Laidlaw is actually remembering accurately, the oddness of the remark might account for its memorability.

It is possible that the problematic examples from Sinclair's and Laidlaw's interviews are simply "performance errors," revealing less-than-total control over the ability to recall actual, or construct appropriate, dialogue. Even if this is the case, the remarkable fact is that this happens so seldom and that the quoted direct speech is generally introduced so effectively that its plausibility is not an issue and the listener accepts the quoted remarks as authentic.

Linguistic Characteristics of Quoted Direct Speech

It was pointed out above that the use of the discourse marker *well* in quoted direct speech corresponds to its use in actual conversation. An even more striking example is the use of *oh* as a turn-taking signal. Schiffrin (1987, 73–101) has pointed out the role of *oh* as a marker of information management, including shifts in speaker orientation. There are 60 examples of *oh* in quoted direct speech in the Ayr interviews, and they all mark a change of speaker. This is particularly important when there is no reporting verb to bracket the quoted speech. The following example comes from Rae's interview:

    (24)   WR145   and the auld yin says to him
           146   "Whaur are you going?"
           147   he says
           148   "I'm going to my work"
           149   "*Oh* you got a job?"
           150   "Aye"
           151   "Whit are you daeing?"
           152   says
           153   "I'm laboring tae a bricklayer in the town"
           154   "*Oh* very good
           155   away and get your bus" he says

In lines 149 and 154 *oh* is turn-initial and marks a change of speaker that is not otherwise signaled. It is remarkable that speakers in quoting direct speech include such items as *oh* and *well*, which have clear discourse functions that would not be contained in a paraphrase of the propositional content of the utterance or in reported speech (*oratio obliqua*). On the other hand, speakers less frequently include the discourse markers *you know* or *I mean*, which refer to hearer and speaker participation. Polanyi points out that the pronouns in *you know* and *I mean*, when used in the course of a narrative, will be taken to refer to the teller and the listener and not to anyone in the story world (1985, 195). This suggests that quoted direct speech has its own properties, distinct from other forms of speech.

One example of this can be seen in the syntactic structure of quoted direct speech. In a comparison between sections containing quoted direct speech and the interviews as a whole there is a considerable difference in the frequency of coordinate clauses, which make up only 10% of the total syntactic units in quoted direct speech, compared with 26% in the interviews as a whole. Since coordinate clauses are particularly important in maintaining the floor and extending a turn, it is perhaps not surprising that they should be less frequent in quoted direct speech. Speakers seldom report themselves or others as having long turns. Instead, there are frequently adjacency pairs such as question and response. This can be seen in the much higher proportion of questions in quoted direct speech (11%), compared with less than 2% in the interviews as a whole. The interactive nature of quoted direct speech is also seen in the high proportion of imperatives (7%), compared with .3% in the remainder of the interviews. Quoted direct speech thus reveals much more interaction than the speech in the rest of the interview.

Conclusion

The use of quoted direct speech adds liveliness to a narrative, provides the possibility of a different perspective, and can give an impression of authentic recollection of an event, though it would be unwise to assume that it necessarily reports the actual words used. By appearing to quote the actual words said, the speaker is giving the listener the opportunity to interpret the statement without the risk of its being distorted by the form in which it is reported. More importantly, the speaker can quote someone else as saying things in a way that presents the speaker in a favorable light, whereas in a summarized form that implication might be lost unless stated explicitly, in which case the speaker might be perceived to be boasting. Like many other features of discourse, it is a rhetorical device. Its widespread effectiveness is a reflection of the skill, which many speakers have, of recreating dramatic dialogue that is appropriate to the protagonists and the scene. As Polanyi has remarked: "Both for literary artists and everyday tellers, the problem of finding a place to stand in order to report the goings on in another world while carrying out one's role as a competent and trustworthy member of society (who cannot know what he or she could not know, for example) is a problem of telling and not the locus of 'verbal art' " (1982, 169). Quoted direct speech allows the speaker an additional stance.

As Abercombie has observed, "monologue" is a very specialized use of spoken language and has "certain distinguishing linguistic, including phonetic, peculiarities" that set it off from other uses of language (1965, 2). The speech in the best of the Ayr interviews approaches close to monologue in the sense that the speakers use extended turns; it is, of course, dialogic throughout, since it is addressed to the interviewer and whoever else happens to be present. As mentioned earlier, in interviewing I tried to say as little as possible, leaving it to the respondent to feel the responsibility of filling in the silence. In the most successful interviews the respondents took advantage of extended turns to tell stories that included quoted direct speech. In the interviews with Gibson, Nicoll, Lang, and Laidlaw more than 10% of the interview consists of quoted direct speech.

This frequent use of speech attributed to others allows the language of the interview to become "polyphonic" in Bakhtin's sense (1973, 4). Instead of a single point of view, the speaker is able to present different perspectives while retaining control of the discourse. To quote Voloshinov again, "Reported speech is speech within speech, utterance within utterance, and at the same time also *speech about speech, utterance about utterance*" (1986, 115, emphasis original).

I have examined the use of quoted direct speech in the Ayr interviews and shown that speakers use quoted direct speech to provide internal or embedded evaluation; to display knowledge of, and attitudes toward, other forms of speech; to make claims about successes or skills indirectly; and to add credibility to a narrative. The use of quoted direct speech, however, is no guarantee of its accuracy, since speakers may quote others as using forms that are improbable in the context. The language used in quoted direct speech differs most from that of the rest of the interview in containing more questions and imperatives and fewer examples of coordinate clauses. This reflects the more interactional nature of the quoted speech and the fact that speakers do not use direct quoted speech as a way of telling extended narratives. Speakers also display an ability to reproduce turn-taking signals in reporting dialogue. In their use of quoted direct speech, speakers reveal in many ways their knowledge about the use of speech within their community.

12

The Individual Interviews

The preceding chapters have shown that there are considerable differences between the social class groups in the use of certain linguistic features. The question that arises is, How reliable is this information? This is not a question about the statistical significance of such numbers as have been presented or about the representativeness of the sample. Rather, it is a much more basic (even simpleminded) question: Do the tapes merely record "interview speech" (in the negative sense used by Wolfson 1976 and Milroy and Milroy 1977), or do they provide the basis for a (partial) description of Ayr speech as used within the community? My own view is clearly the latter; and though I am obviously biased, I believe the kind of evidence presented in earlier chapters supports this position. In this chapter, however, I want to make the case for the interviews as conversations between "intimate strangers" (Albris et al. 1988, 70).

As was pointed out in Chapter 2, the speakers are responsible for raising half the topics discussed on the tapes. For convenience, Table 2–3 is repeated here as Table 12–1. The figures in Table 12–1 show the speakers as a total group to be active participants in the discourse, initiating more than half the speech on the tapes. It should not be thought, however, that the remainder of the speech consisted of short answers to questions. Table 12–2 presents a comparison of topics I raised with those initiated by the other speakers. It can be seen from Table 12–2 that speakers are as talkative on topics that I raised as they are on those they initiate themselves. The measure of syntactic units was chosen for convenience, since that is how the transcripts are coded. Though inexact, it is probably accurate enough for this kind of comparison. It should be pointed out, however, that it is not easy to determine the boundaries of a topic. Grimes gives the commonsense definition: "As long as the speaker continues talking about the same thing, he remains within a single segment of the text at some level of partitioning" (1975, 103). The difficulty is to know

Table 12.1. Proportion of Topics Initiated by Respondents

Respondent	Topics Initiated by Respondent	Proportion of Total Interview Initiated by Respondent
Lower-class	(%)	(%)
Ritchie	13 (4/30)	4
Rae	13 (7/55)	15
Gemmill	23 (23/42)	67
Sinclair	56 (43/77)	50
Lang	77 (72/93)	82
Laidlaw	85 (52/61)	68
All LC	56 (201/358)	59
Middle-class		
Muir	12 (4/34)	12
MacDougall	26 (12/46)	20
Menzies/MacGregor	35 (36/104)	28
Gibson	44 (30/68)	44
Nicoll	78 (63/81)	82
All MC	44 (145/333)	44
All speakers	50 (346/691)	53

Note: Since Menzies and MacGregor were interviewed together, their joint scores indicate the proportion of topics that were not initiated by the interviewer.

what constitutes "talking about the same thing" even in an interview when the interviewer asks a specific question:

(1)   (/)Where were your folk from?(/)
WL006   my—my father belongt tae / Dalry
007   and my mother was / from Annbank
008   but she come to Whitletts
009   when she was very small you see
010   and eh—she went to country service
011   she went to the farms to work you see
012   in fact she ran away fae a farm / Roadend
013   what you called Roadend
014   and / eh her father took her back
       (How old was that?)

Lines 6–9 are clearly a direct response to the question, but are lines 10–14 on the same topic? This is the kind of decision that faces the analyst. There are two criteria for segmentation. One is notation, or what Hinds calls "a unified

Table 12.2. Length of Topics (in Syntactic Units)

Respondent	Interviewer-initiated	Respondent-initiated
Lower-class		
Ritchie	30	9
Rae	15	18
Gemmill	30	51
Sinclair	54	43
Lang	25	34
Laidlaw	43	38
All LC	34	38
Middle-class		
Muir	26	27
Menzies/MacGregor	27	19
MacDougall	40	29
Gibson	24	24
Nicoll	29	37
All MC	29	29
All speakers	31	34

orientation" (1979, 141). The other criterion is to identify discourse boundary markers (Brown and Yule 1983, 69; Van Dijk 1982, 181; Jefferson 1978, 221). Lines 7–14 are about Willie Lang's mother, which gives them a certain unified orientation; but they contain at least four items of information: (1) she came from Annbank, (2) she came to Whitletts when she was very young, (3) she went to work on a farm, and (4) she ran away from a farm. Lines 6–9 also have a unified orientation in that they give the information about Lang's parents requested in the question. Since his mother had come to Whitletts when she was very small, she is really from Whitletts although she had been born in Annbank. Looking at the syntactic organization, the first four clauses are linked by the coordinating conjunctions *and* and *but* and by the subordinating conjunction *when*. Line 9 also ends with the discourse marker *you see*. As was shown in Chapter 10, this is a very frequent item in Lang's interview, occurring no fewer than 124 times. It is most common in clause-final position and seems to function most frequently as some kind of terminal boundary marker. Its occurrence at the end of line 9 is therefore probably not accidental and supports the view that lines 6–9 are a discourse unit.

Line 10 begins with *and eh*. This is not the same *and* as in line 7 but an initial boundary marker (cf. Johnstone 1988). Line 11 also ends with *you see*, marking Lines 10–11 as a unit of some kind. Line 12 begins with another discourse marker—*in fact*—which occurs predominantly in initial position. The section ends (though the topic is not necessarily exhausted) with the interviewer's question after line 14. There are thus formal grounds for arguing that lines 10–11 are a new topic, as are lines 12–14. Since there are also notional grounds for treating them as separate topics, the answer seems clear. On the other hand, lines 6–14 are a response to a simple question and in that sense have "a uniform orientation." Lines 8–9 illustrate what Hurtig (following Schegloff and Sacks 1973) calls *topic shading,* which is "a form of shift which involves the expansion of the domain or scope of the original propositional set" (1977, 95). Lines 10–14 illustrate what Hurtig calls *topic fading,* which "involves the establishment of a new propositional set with a link to either a predicate or argument in an antecedent propositional set" (1977, 96). The distinction, though subtle, seems to fit the sequence quite well (though the difference between expanding the domain and adding a new link may be difficult to determine in many cases). A further problem, as Hurtig points out, is "that topic transitions or boundaries can be marked or unmarked in the surface configuration of the discourse" (1977, 96).

These problems are most acute in dealing with topics initiated by the interviewer, since with rare exceptions the responses are relevant but may diverge in a coherent form (as in Item 1) from the original question. In the topics that are clearly initiated by the respondent it is less complex to keep track of topic shifts. This is because the speakers use a variety of boundary-marking devices to introduce new topics (see Appendix II for details). I found that by dividing up the interview into impressionistic units where the speaker seemed to be "talking about the same thing" and noting how these notional units were initiated, it was usually possible to identify the "speaker's topics"; and I used my questions or remarks as the initial boundaries of "interviewer's topics." Table 12–3 shows the pattern of Ella Laidlaw's interview. It can be seen from Table 12–3 that Ella Laidlaw did not need much encouragement to talk. A very different situation can be seen in Table 12–4, which gives the equivalent representation of the interview with Mary Ritchie. These two interviews with women of similar age and background who lived very close to each other show the two extremes of the interviews collected. The difference between the two interviews will be brought out more clearly in the more detailed discussion below.

The model that seems to provide the best framework for understanding the sequence of topics is Reichman's notion of "context space" (1978, 1985). Reichman claims that

> a discourse is partitioned into a set of hierarchically related constituents, which are only partly defined by the conversational moves they contain. Discourse

Table 12.3. Laidlaw: Topics Elicited and Volunteered

Lines	Topics Elicited	Topics Volunteered
1–40	—	son in U.S., daughter in Canada
41–88	more about son and daughter	—
89–151	family background	—
152–78	—	SON HAD TO LOSE WEIGHT
179–213	growing up then and now	—
214–365	early forms of amusement	—
366–419	relations with parents	—
420–77	children's games	—
478–511	—	a schoolcase of wood
512–96	—	TROUBLE AT SCHOOL
597–669	—	THROWN INTO THE POND
670–705	education then and now	—
706–85	—	PLAYING TRUANT
786–98	left school at sixteen	—
799–809	—	education then and now
810–62	—	ILLITERACY NOW
863–91	—	DEALING WITH ILLITERACY
892–1047	—	GRANDSON'S EDUCATION
1048–77	—	grandson and television
1078–1100	—	used to entertain themselves
1101–37	—	children then and now
1138–43	first job	—
1144–1303	JOB IN CARPET WORKS	—
	[episode in which the husband talks about early work experience]	
1335–1418	bad language nowadays	—
1419–33	—	TWO TEENAGERS TODAY
1434–66	—	HOW SHE GOT INTO TROUBLE
1467–1512	—	social worker's efforts
1513–45	—	teenagers nowadays
1546–53	—	A ROBBERY
1554–79	—	importance of family
1580–1623	—	a good youth leader
1624–51	—	living conditions
1652–76	—	against gossiping
1677–1744	—	language and blue jokes
1745–69	—	daughter's singing

(continued)

Table 12.3. (*Continued*)

Lines	Topics Elicited	Topics Volunteered
1770–88	—	daughter's blindness
1789–1822	—	blind son-in-law
1823–57	—	SON-IN-LAW AND SNP
1858–1914	—	son-in-law's blindness
1915–89	—	BUYING A PRESENT
1990–99	—	blindness in the family
2000–2031	—	her granddaughter
	[narrative by husband]	
2034–78	—	son-in-law's parents
	[narrative by husband]	
2081–2140	—	son-in-law's parents
	[narrative by husband]	
2146–2166	—	son-in-law's parents
	[narrative by husband]	
2167–82	—	A BOAT TRIP
2183–2200	—	gardening
2201–15	—	old age
	[reminscence by husband]	
2222–34	—	questions about interviewer's origins
	[break in tape]	
2235–73	—	reminiscence about father
2274–2350	—	FATHER'S DEATH
2351–64	—	reminiscence about mother
2365–2415	—	MOTHER'S STRICTNESS
2416–62	—	reminiscence about mother
2463–89	—	vandalism at present
2490–2522	—	A PRANK WHEN A CHILD
2523–58	—	more on vandalism today
2559–89	—	CASE OF VANDALISM TODAY
2590–2640	—	racial differences
2641–55	—	working men today
2656–84	—	UNIONS TODAY
2685–2739	—	A STRIKE IN THE PAST
2740–60	—	problems in the old days

Note: Capitals indicate narratives.

9 topics initiated by interviewer = 15%

52 topics initiated by respondent = 85%

Proportion of interview initiated by respondent = 68%

Average length of topic = 45 lines

Table 12.4. Ritchie: Topics Elicited and Volunteered

Lines	Topics Elicited	Topics Volunteered
1–8	—	about interviewer
9–71	background information	—
72–84	—	misses her garden
85–131	life then and now	—
132–62	husband's life	—
163–94	childhood memories	—
195–213	differences between then and now	—
214–338	children's games	—
339–88	schooling	—
389–93	Ayr people not snobbish	—
394–450	work after leaving school	—
451–77	work after getting married	—
478–85	meeting husband	—
486–95	going to dances	—
496–507	holidays	—
508–30	changes in Whitletts	—
531–44	World War II	—
545–50	children's lives	—
551–67	superstition and luck	—
568–85	a quiet life	—
586–90	they don't work as hard now	—
591–603	Gray's factory	—
604–14	—	stamp works may close
615–98	Community Centre	—
699–702	—	going to football
703–39	holidays	—
740–42	likes living here	—
743–66	Scots vs. English	—
767–818	old words and expressions	—
819–21	no second chance at seventy	—

26 topics initiated by interviewer = 87%

4 topics initiated by respondent = 13%

Proportion of interview initiated by respondent = 4%

Average length of topic = 27 lines

participants "package" pieces of discourse into these separate units and selectively bring these units in and out of the foreground of attention in the generation and interpretation of subsequent utterances. The fundamental unit of discourse processing—the constituent hierarchically related to other discourse constituents and brought in and out of focus in a discourse—we call a *context space*. (1985, 24)

In the Ayr interviews, what Reichman calls the "controlling context space" was the respondent's recollections of life at an earlier time. This is the issue to which respondents return at various points during the interview, without there being any sense of loss of direction. In this way the speakers maintain "coherence" as well as "cohesion" (Tannen 1984, xiv) in their contribution to the discourse.

There is another aspect to the transactional relationship, namely, turn taking. As mentioned previously, in my role as interviewer I tried to yield the floor whenever possible. In other words, I tried to mimimize the question-and-answer format, which often results in the sequence of short turns deplored by Wolfson and the Milroys. The extent to which this tactic succeeded can be seen in Table 12–5. The variation in turn length that these figures conceal will be shown in considering the individual interviews, but it is clear from Table 12–5 that all the respondents provide more than minimal responses to questions. Although length of turn is not generally considered the critical variable

Table 12.5. Average Length of Turn in Syntactic Units

Respondent	Length of Turn
Lower-class	
Rae	8.0
Ritchie	8.6
Laidlaw	21.7
Gemmill	29.7
Lang	35.6
Sinclair	60.2
Middle-class	
Menzies	5.2
MacGregor	9.7
Muir	12.9
MacDougall	17.1
Gibson	23.9
Nicoll	37.8

in judging quality of speech, when long turns occur in an interview situation, it is usually an indication that the respondent is in control. This is obviously the case in extended narratives but can also be true of detailed descriptions or reminiscences. The relationship between turn and topic is a complex one, since a topic may be sustained over several turns and since (conversely) within a speaker's turn there may be topic shifts.

There is, however, one confounding factor in Table 12–5. The interview with Menzies and MacGregor was a group session, and Laidlaw's husband was present for most of the interview with her. In these sessions there are, consequently, many short turns that are responses to remarks other than those initiated by the interviewer. Such turns do not have the stigmatized character of the question-and-answer format of "interview style." The account of the individual interviews will show the number of extended turns for each speaker, and this may give a better indication of the kind of language to be found in the interviews. The figures in Table 12–5 also conceal the shifting character of the exchanges. Although it is not true in all cases, there is a tendency for the turns to be shorter at the beginning of the interview, rise to a peak of longer turns at some later point in the interview, and then revert to somewhat shorter turns as the interview winds down. This might appear to be a natural progression; but as will be seen below, it is not a universal characteristic of the interviews.

The Individual Interviews in Detail

The first set of interviews to be considered are those where the influence of the interview situation is most obvious. These are the five with the shortest average turns: Muir, Menzies, and MacGregor among the middle-class speakers and Rae and Ritchie in the lower-class group.

Muir

30 topics initiated by the interviewer = 88%
4 topics initiated by the respondent = 12%
Proportion of interview initiated by respondent = 12%
Average length of topic = 26 syntactic units
Number of narratives = 0

This was the shortest of the middle-class interviews (4,573 words) and began with the least promising opening:

(2) IM001 are we on tape?
           (Yes do you mind?)

002 I'd like
003 to know more about it
004 yes I do mind
005 I mean I want
006 to know more about it
007 because it seems the strangest thing
008 is it a political thing?
    (No no it's not political
    I teach at a private university
    and I'm interested in the way people speak.
    I've just come from talking with Hugh Gemmill)
009 I know
010 he comes here
011 I know Hugh fine
    (I'm interested in the fact that you two can communicate
    with each other and yet you speak very differently)
012 yes
013 how do you know
014 we can talk?
015 did he tell you that?
    (No he didn't mention it but I'm assuming—)
016 I see
017 he's a nice chap

While this exchange has the form of question and answer, it is Muir who is doing the asking. It is clear from the start that Muir is not totally passive in the situation; nor is he intimidated by the presence of the recorder, despite his first question. With hindsight, it is apparent that neither I nor my network contact had prepared him for the situation. I had assumed that he knew that I wanted to interview him and for what purpose. Considering the miscommunication, it is surprising that the interview turned out as well as did. Perhaps it was because of his sense of noblesse oblige.

Muir comes at the upper end of the social class spectrum in the middle-class group. He was the only one to have been sent away to a private school for his secondary education. He had been a successful businessman. He had decided against the law as a profession, though he had started his apprenticeship before World War II, planning to join his father's firm. Once he had satisfied himself that it wasn't "a political thing," he was willing enough to be interviewed; but it was apparent throughout that he did not like talking about himself and from time to time he would say things such as "Anyway you don't want to hear any more about that" or "I don't want to get talking

politics," and questions that could be considered in any way personal tended
to be given rather short answers. His response to the "danger of death"
question (Labov 1966a) was to refer only to his wartime service; and when I
pushed him, his response was as in (3):

> (3)   IM790   oh I suppose
>   791   you could say
>   792   I was
>   793   yes I would say so
>   794   but I don't mean
>   795   that I had a very hard war
>   796   I'm not complaining about it
>   797   but I think
>   798   there would be occasions
>   799   when I was distinctly frightened

It is difficult to tell whether this is modesty or because there really was
nothing very exciting to report. In the circumstances of the interview I felt that
it would not be polite to probe further.

Less than half of Muir's turns (42%) are longer than 10 lines. Most of his
remarks provide factual information or explanations; there are no true narratives, though about a third of the interview consists of reminiscences. His
longest turn (119 lines) describes his career in business.

Muir's interview has the lowest proportion of coordinate clauses in the
middle-class sample, a high proportion of subordinate noun clauses and of
that-deletion, and a high frequency of infinitives. Muir has also the highest
proportion of nonrestrictive relative clauses in the whole sample. Since these
are characteristics of the middle-class group as a whole in comparison with
the lower-class group, it is appropriate that Muir's interview should exhibit
them strongly. Muir's interview, however, also contains the highest frequency
of discourse markers in the middle-class sample. In particular, he has by far
the highest frequency of *you know* among the middle-class speakers. The use
of discourse markers probably is an indication that despite his general re-
sistance to the interview and line of questioning, Muir's speech is not greatly
affected by his awareness of the tape recorder. In addition to talking about his
business, the other topic that brought extended responses from Muir was
sport. Here is his response to the question "What sports did you play?":

> (4)   IM394   you mean when I was older?
>   395   when I was able to—
>   396   well—well what we played at school
>   397   was rugby and cricket—athletics

398 I played squash at school
399 I played fives
400 I played tennis—and—
401 but principally—eh—rugby and cricket
(How long did you go on playing after you left school?)
402 well of course I was interrupted very much by the war—
em—just at a time
403 when I would have enjoyed them most probably—from
nine—nineteen until twenty-six
404 a bit of a gap
405 although I was able
406 to play a little of each during the war
407 but—very—in—not—not continually—nothing like it
408 well when I came back from the war—em
409 it was difficult
410 to [coughs] play rugby initially
411 because I—I was in a shop and—eh—you know
412 but—em—I was able
413 to start
414 playing
415 I think
416 it was in forty-six
417 I don't remember
418 and I used to play rugby
419 used to go to the business in the morning
420 cycle out to the old racecourse
421 have my lunch
422 and cycle back to business
423 never thought anything about it
424 well I did that
425 till I was thirty and then after—
426 it was becoming a bit of a bind
427 because—em—obviously you're older
428 and I suppose
429 being on your feet all morning
430 and then playing rugby all afternoon
431 it became less attractive
432 and when I got a kick
433 instead of you know being a bit bruised and sore on
Monday
435 and being all right on Tuesday

436   it was taking to Thursday
437   before it got better
438   and I thought
439   "Well there's not much point
440   in playing
441   if you don't enjoy it so much"
442   I stopped that then
443   and—eh—but I played cricket
444   till I was about forty
445   and I stopped it principally because of business actually
      not because of health or anything like that or not be-
      cause—
446   I stopped it principally because of
447   getting more involved in business
448   and it was—
449   I was finding it difficult you know
450   I was away on a Saturday
451   Sunday was fiddling and that kind of thing
452   you know I just sat Sunday night and things
453   that's
454   why I stopped it really
455   and then I started
456   playing
457   I played golf then
458   which I could kind of determine—more control
459   when I played you know
460   to suit—to suit
461   and that was—
462   and then I—I fished a bit
463   but principally it was golf after that

Note the use of *well* in lines 396, 402, 408, 424, and 439. This is almost a textbook illustration of points mentioned in Chapter 10. The example in line 396 illustrates Schiffrin's category of an "embedded referent/response pair," line 402 illustrates an indirect response to a WH-question, the instances in lines 408 and 424 indicate transitions in the discourse, and line 439 shows the use of *well* in quoted direct speech even though the speech is his own. There are also four examples of *you know,* and one of *actually.* There are the hedges *a little of* (406), *a bit of* (404), *a bit* (433, 462), *kind of* (458), and the "loosely interpreted" quantifier *all* (429–30). There is a high frequency of adverbs: *principally* (401, 445, 446 463), *very much* (402), *most probably* (403), *continually* (407), *initially* (410), *obviously* (427), and *really* (454). Notice also the sequence of subjectless verbs in lines 419–23. This is characteristic of a certain laconic style that tends to depersonalize the description. Of a similar kind are the clichés *a bit of a bind* (426) and *there's not much point in* (439).

The extract in (4) is typical of the kind of extended answer Muir often gave to my questions. It is not very fluent, but the rather staccato delivery is not uncharacteristic of upper-middle-class male speakers of his generation in the West of Scotland. While the interview cannot compare with several others in terms of quantity and variety, there is no reason to believe that it gives a misleading impression of Muir's speech in conversation with a stranger.

MacGregor and Menzies

68 topics initiated by interviewer = 65%
36 topics initiated by respondents = 35%
Proportion of interview initiated by respondents = 28%
Average length of topic = 24 syntactic units
Number of narratives = 9 (all by MacGregor)
Number of words = 5,110 (Menzies) and 7,673 (MacGregor)

This was an interview that I felt at the time ought to have come off better than it did. Since there were three people present—MacGregor, his wife, and Nan Menzies—I had hoped that it would turn into a self-sustaining discussion with the interviewer playing a relatively minor part. This did not really happen, although there were times when the three of them did talk to each other rather than to me. For the most part, however, MacGregor and Menzies addressed their responses to me rather than to each other; and the presence of the other may even have had an inhibiting effect, since I was asking them to talk personally about a background they had shared. They were generally too polite to contradict each other, though there were a few examples later in the interview. To some extent the fault was mine for accepting a physical setup that emphasized distance. The interview took place in the MacGregors' comfortable drawing room with the chairs at a respectable distance from each other. I was seated too far away from both MacGregor and Menzies, which would have been less of a problem if the discussion had been the self-sustaining kind but was unfortunate because of the major role I had to play as interviewer. In the course of the interview I asked over a hundred questions, more than in any other interview. Partly, this was because I repeated my questions for MacGregor and Menzies; but it had the effect of constantly reinforcing the notion of the speech event as an interview and the physical distance was almost symbolic of a communication gulf between us. This was all the more annoying in that MacGregor and Menzies are the kind of people I grew up with (and they knew that as well as I did); but somehow the roles of interviewer and interviewed proved to be the strongest determinants in the interview. Contrary to my original impression, however, the tape of the interview shows that at least in MacGregor's case the interview had provided a substantial amount of continuous speech.

Although less than a third (29%) of his turns are longer than 10 lines, MacGregor has eight extended turns of more than 40 lines, the longest being 87 lines. He tended to talk rather softly and slowly, and I found myself increasing the volume and speed of my own speech slightly in the hope that he would respond in similar fashion; but he maintained his own style. At the beginning of the interview, his speech has many false starts and filled pauses, as illustrated in (5):

(5)   JM029   my father was originally from Prestwick
       030   but he'd lived in Ayr
       031   well he was a—
       032   he ran away from home
       033   when he was very young—twelve or something
       034   and—and—uh—eventually joined a whaler
       035   he wo—he worked on a whaler for a time
       036   and—uh—somewhere on his travels he must have qualified as a—an electrician
       037   and he came back here
       038   and—uh—lived in—eh—Fotheringham Road
       039   to begin with for a while

But MacGregor also gave fairly full answers to my questions and approximately a quarter of his interview consists of narrative. He is the third highest in the middle-class group in his use of quoted direct speech. In his use of syntactic structures he is close to the average for the middle-class group, though he is low in his use of discourse markers and has the lowest proportion of *that*-deletion in the group. He also has the highest frequency of WH–relative markers. These latter two characteristics may reflect his profession as a schoolteacher. The extract in (6) is a good example of his narrative style in response to that often-effective question, "What did you do when you left school?":

(6)   JM647   I went to Jordanhill
       648   as a matter of fact I was sitting in—in a classroom
       649   when I was in fifth year

650 and—eh—a girl wee Jean Robertson said
651 that she had sent away her application for Dunfermline
652 and that you didn't need
653 to have your Highers
654 to get in
655 as long as you were presented
656 and of course I had been doing no work at all
657 and I wasn't going to be presented for my Highers in the fifth year first of all
658 so I asked the teacher
659 if I could go and see the rector
660 and I spoke to one of the joint rectors
661 I said
662 "I've just discovered
663 I don't need my Highers for Jordanhill
664 but I've got
665 to be presented
666 do you think
667 I could be presented?"
668 and he said
669 "Oh no MacGregor
670 we don't do things like that in Ayr Academy
671 and if you had worked
672 you would have been presented
673 we can't possibly present you"
674 and just as I was going out
675 I must have been looking pretty crestfallen
676 when the other rector came in
677 and he had more of a soft spot for games—eh—Jamieson
678 and he said
679 "Well MacGregor what's wrong?"
680 and I said
681 I was down
682 to see
683 if I could be presented
684 "Oh wait a minute now
685 wait a minute"
686 and he and Willie Dick went away over to the window
687 and had a—had a discussion
688 and then they came back

689   and Mr Dick said
690   "If we present you
691   will you work?"
692   and I said
693   "Oh yes"
694   so they presented me
695   and I had extra—extra maths from Hughie and extra
        English from Willie Dick
696   I scraped a pass in English and something else
697   so I got into Jordanhill

This is a well-told story with good use of both reported speech and quoted direct speech. There is no use of the so-called conversational historical present tense (Wolfson 1982), which is commonly found in narratives with more action in them and which MacGregor uses in a later narrative about an accident. The syntax is varied with the greater part of the narrative being carried by coordinate clauses and quoted direct speech. The discourse markers *oh* and *well* are used effectively to bracket quoted direct speech. The telling is fluent, with none of roughness of the extract in (5). Despite the stiffness of the interview situation, MacGregor turns out to have been more responsive than my original impression of the interview suggested. Perhaps the softness with which he sometimes speaks suggests less involvement than turns out to have been the case. Not only in the narrative passages but also in places where he expresses strong opinions on topics such as comprehensive schools and the welfare state he can be quite a forceful speaker, and the interview probably gives a reasonable sample of his speech.

The interview was less successful in eliciting speech from Menzies. She contributes only a little over a third of the speech on the tape, and there are no narratives; but this reticence may not be solely the result of the interview situation. At no time did she give the impression that she was in the habit of telling stories; and although there were many places in her remarks where one might have expected her to illustrate the point with a narrative, an anecdote, or a reminiscence, she never does so. The extract in (7) is an example:

(7)   NM284   and we cycled
        285   and I remember
        286   learning
        287   to ride on a back-pedaling bicycle
        288   the brake was by
        289   pedaling backwards
        290   I had many a tumble on that
        291   before I mastered it

Many people would have had a particular memory of "a tumble" at this point; but Menzies' memory does not seem to focus on dramatic events, and even MacGregor's two good examples of "dangerous situations" did not inspire her to respond other than "I don't think I can match any of these colorful—can't think of anything offhand." Yet there is little sign of nervousness in her speech. Although she at one point says she does not like the sound of her own voice on tape, there is no suggestion that she is uncomfortable in the presence of the tape recorder. There is little change throughout the interview. The average length of her turns does not vary throughout the interview. Only 14% of her turns are 10 lines or longer, and the longest is only 40 lines.

It seems as if Menzies is content to speak quite simply and undemonstratively. Just over half of the clauses in Menzies' interview are main clauses, the highest proportion in the middle-class group. The frequency of nonfinites is low. There is little embedding. On the other hand, the frequency of discourse markers is the second highest in the middle-class group; and she shares with Muir the highest percentage of *that*-deletion. Perhaps not surprisingly since she had trained at "dough school" (i.e., a domestic science college), the place where she becomes most eloquent is in speaking of food:

(8)	NM1379	that is skirlie
	JM1380	I've never had that
	NM1381	oh well Jock it's awfully tasty
	JM1382	I've heard of it
	1383	but I've never tasted it
	NM1384	if you even want
	1385	to put say a chopped onion and some dried fruit in it
	1386	it makes a—like a white pudding
	1387	that you buy in the butcher
	1388	and slice that sort of thing
	1389	we used to—
	1390	one of my favorites was a haggis a sweet haggis
	1391	which mother made
	1392	sheep's stomach bags
	1393	which of course you could get from the butcher
	1394	it was always a big one and a little one of about that size
	1395	and she made this marvelous mixture of oatmeal beef suet chopped seasoning salt and pepper and usually currants
	1396	and this was all put into this—these two bags
	1397	they were all sterilized and clean and scraped and sewn with a big darning needle and a bit of thread

1398   and then they were boiled
1399   and then after they were cooked
1400   they were put in this warm cupboard
1401   we had where—
1402   the kitchen range and the pipes went through a nice cupboard
1403   where you had biscuit tins and all sorts of things
1404   that kept close to it
1405   and this was kept in there in a big ashet
1406   and then for high tea—
1407   it became quite hard
1408   the skin was crisp and hard
1409   and you could cut it with a good knife you see
1410   cut it in a slice
1411   and you laid that on a plate under the grill
1412   and just gave it a wee while
1413   and really it was a gorgeous thing
1414   and it really was very—
1415   I tried
1416   to make it since
1417   but you can't get these things
1418   to make it in

This is the kind of recollection Horvath (1987) calls a "reminiscence," where the reference is to a generalized sequence, not to a specific occasion located in narrative time. Like most of Menzies' language in the interview, it is syntactically simple, with about one-third main clauses and less than 25% subordination; but it is full of evaluative qualifiers, *awfully tasty* (1381), *this marvelous mixture* (1395), *a nice cupboard* (1402), *a good knife* (1409), and *a gorgeous thing* (1413). It is very fluent and is addressed as much to the MacGregors in this instance as it is to the interviewer, since John MacGregor had mentioned that he had never tasted skirlie.

I was disappointed in the interview with MacGregor and Menzies at the time because I had expected a more interactive session, with topics arising spontaneously. That did not happen the way I had hoped. Yet on analyzing the tape carefully and listening to kind of speech recorded, I am prepared to believe that it may give a fairly accurate reflection of their language in everyday situations. Probably neither Menzies or MacGregor will ever be boisterous or dramatic in company, and the restrained nature of their recorded speech may be a more representative sample of their everyday language than I had originally thought.

Ritchie

26 topics initiated by interviewer = 87%
4 topics initiated by respondent = 13%
Proportion of interview initiated by respondent = 4%
Average length of topic = 27 syntactic units
Narratives = 0
Length = 4,373 words

In an earlier paper (Macaulay 1984) I used Ritchie as an illustration of someone who was not easy to interview. To this day I am not sure how much was my fault. I certainly tried hard enough, perhaps too hard. I asked a total of 78 questions in the course of an interview of 4,373 words, which works out at just over 50 words a response. While 41% of her turns are 10 lines or longer, only 9% are 20 lines or more, and the longest turn is only 28 lines. The length of her turns remains fairly constant throughout the interview, although they are slightly longer toward the end. It is unlikely, in the circumstances, that the interview gives good evidence of her everyday speech. Her use of a monophthong for (au) almost doubles from the first half of the interview (31%) to the second half (52%) showing some adaptation to the situation. She uses very little /e/ for (e), and she consistently has some form of /a/ for (a). Yet her speech in no sense could be mistaken for middle-class speech. She has considerable vocalization of /l/, and she has the highest frequency of non-agreement of subject and verb of all the lower-class speakers. She uses the negative clitic *–nae, no* for *not, thae, yin,* and *wan.* The extract in (9) gives a good idea of the dynamics of the interview:

(9)		(What did you do when you left school?)
	MR394	eh I went
	395	to work in the factory
		(How old were you then?)
	396	well no just right away
	397	it was aboot two or three month
	398	before I went
		(How old were you then?)
	399	eh—fourteen—fourteen and a half by the time
	400	I started
	401	working
	402	that the age
	403	for leaving—fourteen
		(What kind of job was it?)
	404	eh—carpet weaving

405   it was—it suited—eh it suited me
406   my mother wasnae strong
407   and eh you had er—you only worked a half day on a
      Saturday
408   and that let me
409   get home
410   to do the heavy work you know
411   and just so as my mother just had light work
412   to do all week after it
413   but—eh—I wasnae in love with it either
414   I hated thon big gate
415   clanking at the back of me
416   I felt
417   as if I was in a prison
      (Was it hard work?)
418   oh it was hard work oh aye
      (Did it take you long to learn?)
419   no
420   oh you've
421   to serve your time at it
422   learning
423   you've
424   to serve your time—aboot two year
425   serving your time
426   seven and six a week
427   seven and six a week
428   when you think on it
      (Did you have big machines?)
429   oh aye big—eh—big looms
430   there were two of us on it
431   two of us on the loom one one each end
432   and we just went up
433   and met one another
434   and then come back doon
435   oh the—working looms was heavy
      (How many hours did you work?)
436   eh quarter-to-eight to half-past-five
437   and hour off at dinner time
      (That's a heavy day)
438   it was a long day
439   and then you'd

440   to walk out the road fae—oot to Whitletts [laughs]
441   as soon as I come home
442   I was starving
443   my mother had my dinner
444   sitting
445   and waiting me the end of the table [laughs]
     (How long did you work there?)
446   I worked
447   tae I got married
448   I got married
449   when I was twenty-two

In this interview the question "What did you do when you left school?" failed to perform its magic and instead of the extended responses that I got in some of the other interviews, I got a straightforward answer: "I went to work in the factory." It fulfills the Gricean maxims, but it was not what I had hoped for. Eight more questions and a comment in the next 50 lines serve little better. There are some longer responses in the interview, but at no time does Ritchie really start talking to me as if she were enjoying the experience. Despite this, Ritchie manages to convey why she "wasnae in love" with her work or with "thon big gate clanking at the back of me . . . as if I was in a prison"; and she makes effective use of repetition (Tannen 1989). This is the interview that illustrates the kind of language categorized by the Milroys as "interview style." Although her speech in no way approaches middle-class norms, the evidence from Ritchie's interview as an example of lower-class speech must be treated with more caution than that from the other interviews.

Rae

48 topics initiated by interviewer = 87%
7 topics initiated by respondent = 13%
Proportion of interview initiated by respondent = 15%
Average length of topic = 16 syntactic units
Narratives = 6
Length = 5,030 words

This interview was a disappointment for very different reasons. I had had high expectations of it because Sinclair had told me what a great storyteller Rae was, and he had set up the interview for me. It took place about two weeks after the interview with Sinclair. Some of the blame must rest with me because I had expected it to be one of these easy affairs like the interviews with Lang, Laidlaw, Nicoll, and Sinclair himself, where essentially my only task was to prompt or probe if the flow of talk seemed to be slackening off. As a result, I had not prepared myself as well as I might have done, either mentally or by making sure that I had the right questions ready. It was not that Rae was unfriendly or uncooperative, but he did not seem completely at ease in the situation. Of course, I have no idea what Sinclair said to him when he arranged the interview.

Rae demonstrated his narrative skills in a few examples; but he did not take the initiative, as can be seen in the fact that I asked 76 questions in an interview of only 5,030 words. The average length of his turns is less than a quarter of those in Lang's interview, for example. Less than a third (30%) of his turns are 10 lines or more, and the longest is 54 lines. Rae had a reputation as a public performer, and I had expected that this skill would translate into entertaining personal reminiscence; but it seemed that one person was not a sufficient audience to bring out his rhetorical skills. (There is also the fact that his public performances generally took place in gatherings where alcohol was served; our interview took place in the sobriety of an afternoon in his house.)

All Rae's narratives fall into the category of anecdotes about other people; he did not tell stories of personal reminiscence. An example is given in (10):

```
(10)              (Did people tell stories?)
        WR659     oh aye
           660    I mind once this auld yin telt me
           661    during the wan of the miners' strikes ken years ago
           662    they were away
           663    walking
           664    that's aw
           665    they could dae
           666    they had nae money ken
           667    they were walking away roon by the airport there ken
           668    and this yin hit a good idea
           669    says
           670    "We'll go to Troon"
           671    ken where Troon is?
           672    says
           673    "There's a lot o money in Troon" he says
           674    "We'll collect for Tarbolton Silver Baun"
           675    that was a nonexistent baun of course ken [laughs]
           676    so the two o them—
           677    this auld yin was the best versed of the lot
           678    he says
           679    "Wait here tae I come oot
           680    don't make yersel seen
```

681 I'll gae intae her"
682 so he goes in the gate
683 there was a wee gate
684 and then up to the door
685 rings the bell ken
686 rings it
687 and an auld wuman come oot wi wan o them hearing horns
688 mind ken
689 "We're collecting for the Tarbolton Silver Baun missus"
690 "Whit son?"
691 "WE'RE COLLECTING FOR THE TARBOLTON SILVER BAUN"
692 "Cannae hear you son"
693 and he was roaring it oot
694 "Ach away boys
695 bugger you missus" he says
696 and he walks away
697 and the auld yin shouts
698 "Shut the gate son"
699 "Ah tae hell wi you and the gate" says she—says he
700 and the auld yin says
701 "And to hell wi you and your silver baun" ken [laughs]
702 ah you got some wags

As this extract shows, Rae is a very consistent speaker exhibiting all the essential characteristics of lower-class speech. Although Rae did not open up the way some of the other speakers did, there is no reason to believe that after the first few minutes, in which there are a few diphthongal forms of (au), his speech was affected greatly by the interview situation. The forms *auld, yin, telt, wan, ken, aw, dae, nae, roon, baun, oot, yersel, gae, intae, wuman, whit,* and *cannae* in Item 10 are "robust" features of lower-class speech; and the use of the word *bugger* suggests that he was not inhibited by the presence of the tape recorder. Rae's interview is disappointing only in terms of the quantity of speech produced. Like Muir at the other end of the social continuum, his interview probably gives a fairly accurate indication of his speech in a number of everyday situations.

The MacDougall and Gibson interviews were problematic. These are the two best-educated among the respondents, and it might have been expected

that they would be relatively at ease with a middle-class academic interviewer.
The reality was not so simple.

MacDougall

34 topics initiated by interviewer = 74%
12 topics initiated by respondent = 26%
Proportion of interview initiated by respondent = 20%
Average length of topic = 37 syntactic units
Narratives = 1
Length = 9,764 words

MacDougall was a network contact who was supposed to put me in touch
with others and probably had not expected to be interviewed himself. I had
had difficulty in making contact with him and managed to do so only in the
evening before I was leaving Ayr. He was the respondent who knew most
about my goals in the study and knew that my primary interest was language;
and he had done a certain amount of reading in linguistics. In the circumstances, it may not be surprising that he was somewhat self-conscious. He was
perfectly willing to be interviewed, but his uneasiness persisted well into the
interview. Almost at the end of the interview he remarked:

(11)   AM1607   also I think
        1608   an interesting point would be
        1609   to look at the—interview
        1610   because—
        1611   as it goes on
        1612   because I feel much more relaxed now
        1613   chatting to you
        1614   I'm chatting much more normally
        1615   at the beginning I felt
        1616   it was a bit more stilted
        1617   I'm sure
        1618   that will come through on the tape
        1619   once you play it over

He was right in his perception. The early part of the transcript notes quite
frequently "[clears throat]" and he often apologizes for this. It is, of course,
catching; and the interviewer also clears his throat from time to time but does
not apologize. Despite this slight awkwardness on the part of both the interviewed and the interviewer, this is the second-longest of the middle-class
interviews; and MacDougall's turns are consistently longer than those of
Muir, MacGregor, and Menzies. More than half (52%) are 10 lines or longer

and 13% are more than 40 lines long. There is, however, only one very short narrative and almost no quoted direct speech. Syntactically, the language in MacDougall's interview is the most complex of the whole sample, with by far the highest proportion of subordinate clauses, including noun clauses and the group of adverbial clauses that are more frequent in the middle-class interviews (see Table 8–17); and he has the highest frequency of infinitives (17.9 per thousand words). He has a very low proportion of discourse markers, though about a quarter of them are *you know*. His overall percentage of glottal stops (11.4%) is third-highest in the middle-class group (after Gibson and Nicoll), and this may be a consequence of his lower-class origin. Although there are virtually no narratives, MacDougall gives very full and thoughtful answers to questions. The extract in (12) is his response to the question "Why would you want to settle here rather than elsewhere?":

(12)	AM910	well I quite like this environment
	911	I like the people here
	912	and I like the countryside
	913	and I like the attitudes of people
	914	because I found
	915	one—one problem with say Germany or Oxford was
	916	that there was a certain amount of [pause] unreality in Oxford
	917	in that the academics were really a bit isolated from the rest of the community
	918	and many of them felt
	919	that this was the whole point of
	920	living
	921	to solve their own particular research problems
	922	and nothing else was really all that important
	923	and they tend
	924	to live in this sort of ivory tower atmosphere
	925	although obviously with a generalization like that you know
	926	there were many exceptions
	927	and there were many sort of—sort of normal people
	928	put normal in inverted commas
	929	and in Germany there was a—[pause] a kind of Prussian attitude in especially in the North of Germany
	930	that made it difficult
	931	to get
	932	to know people
	933	I think

934   people tend
935   to keep to themselves a lot
936   or keep to their own family group
937   and there was a certain aggressiveness and arrogance of—arrogance of wealth or arrogance of intellectual status or arrogance of job status
938   which I didn't really like an awful lot
939   and whereas in the West of Scotland I think
940   it's much easier
941   to be accepted for the type of person
942   that you are rather than
943   how much money
944   you've got
945   or what your job is
946   or what Highers
947   you've got
948   so that it's—it's much more down-to-earth and—human
949   those ways
950   and al—another thing
951   which is difficult
952   to explain in—in words
953   is th—th—the kind of humor of the West of Scotland
954   is I think fairly unique
955   I don't think
956   it was just the language in Germany
957   which made me
958   not notice that
959   I think
960   the North Germans tend
961   to be a wee bit humorless
962   although then that's not really fair
963   I think
964   that their humor is different
965   I don't think
966   anyone's humorless
967   but I quite like the—the sort of intellectual atmosphere round here—not in a snobbish sense but just in—in—you know
968   I get on better with people here

Although there are a few hesitations and repetitions in this passage it is on the whole fairly free-flowing. This is achieved by a combination of a high degree of subordination (52%) and regular use of coordinate clauses introduced by *and* (23%). The proportion of subordinate noun clauses is high, as is the frequency of infinitives. This is the most complex syntax in any of the interviews. Once again, there are frequent adverbs and hedges: *quite* (910, 973), *a certain* (916, 940), *really* (918, 924, 943, 968), *particular* (923), *sort of* (926, 929, 973), *obviously* (927), *kind of* (931, 958), *especially* (932), *the type of* (946), *much more* (953), *fairly* (960), *a wee bit* (967), and *just* (975). This effect is reinforced by the many qualifications of statements: "there were many exceptions" (926), "put normal in inverted commas" (928), "especially in the North of Germany" (931), "arrogance of wealth or arrogance of intellectual status or arrogance of job status" (937), "although that's not really fair" (962), "not in a snobbish sense" (967). Such hedging and qualifying is not simply a feature of middle-class speech, it is a particular characteristic of academic discourse; thus, it is not surprising that MacDougall should manifest it so obviously.

This cannot be considered a completely successful interview. It gives a very one-sided view of MacDougall's language. Apart from the nervousness at the beginning of the interview, however, it gives a fairly good example of one style of his speech, namely how he talks as a professional on academic subjects to an academic interviewer. The interview on the whole is "a wee bit humorless." Under different circumstances, there would have been greater variety in intonation, rhythm, and loudness; and presumably there would have been more examples of discourse markers and generally less complex syntax. MacDougall's interview shows that variety is not solely a function of length and that ease in the interview situation is not simply a matter of similarity of background and interests.

Gibson

38 topics initiated by interviewer = 56%
30 topics initiated by respondent = 44%
Proportion of interview initiated by respondent = 44%
Average length of topic = 24 syntactic units
Number of narratives = 6
Length = 8,510 words

As was mentioned earlier, Gibson is one of the better-educated respondents. In addition, his work in the theater has given him a confidence in his ability to mimic other forms of speech. Despite this, he is not a fluent speaker himself. His is the interview most characterized by hesitation, false starts, filled pauses, repetitions, and repairs, particularly in the early part of the interview. In the following extract he is explaining why he was not chosen for parts in the Christmas plays while at primary school:

(13)	WG071	I was never chosen in this
	072	because I was—somewhat bad reader—
	073	eh—reading
	074	by that I mean
	075	speaking
		(Reading aloud)
	076	reading aloud
	077	I was always a bit hesit—
	078	I could—
	079	even tod—even today—I—I—I don't know
	080	that I can get my tongue
	081	to reproduce
	082	what my eye sees—ah—quickly enough

As will be illustrated later, Gibson can be quite fluent—even eloquent—in telling a story; but generally, his style of discourse in the interview is jerky and anything but smooth-flowing.

Gibson's interview is exactly at the median point in length of the middle-class interviews, being twice as long as Muir's and half the length of Nicoll's. The average length of his turns (23.9 lines) is second only to Nicoll's among the middle-class speakers. More than two-thirds (68%) of his turns are 10 lines or longer, with the longest being 106 lines long. After a relatively slow start, the average length of his turns rises in the second half of the interview to over 45 lines. He has six narratives, making up approximately 25% of the interview; and with 10.4% quoted direct speech, he is second-highest in the middle-class group, after Nicoll. There is, accordingly, a substantial and varied sample of his speech to analyze.

In the distribution of main, coordinate, and subordinate clauses; in the proportion of subordinate noun clauses; in the percentage of *that*-deletion; in the distribution of subordinate adverbial clauses; and in the frequency of infinitives, gerunds, and present participles—Gibson's interview lies at about the average for the middle-class group. Syntactically, therefore, the language of Gibson's interview seems to be characteristic of middle-class speech in Ayr, as represented by the sample.

In his use of discourse markers, however, Gibson turns out to be much less typical of the middle-class group as a whole. The frequency of discourse markers in Gibson's interview is only 10 per thousand words and only 6% of the syntactic units contain a discourse marker. If Östman (1982, 170) is correct that the use of such markers is characteristic of impromptu speech, there may be a connection between the dysfluency and the low frequency of these markers.

Gibson is the middle-class speaker with the highest proportion of glottal stops (35.7%) with a frequency of over 50% before a pause followed by a vowel in the next word. This is three times as high as Nicoll, who is the next-most-frequent user of glottal stops in the middle-class group. As pointed out in the previous chapter, Gibson claimed to speak in a wide variety of styles "down to speaking quite Scots." Apart from the narrative sections where he is imitating lower-class speakers, Gibson does not speak "quite Scots" in the interview. In syntax and vocabulary his language in the interview is typical of the middle-class group, and no other aspects of pronunciation show characteristics of lower-class speech. The high frequency of glottal stops, therefore, suggests that despite the dysfluencies and low frequency of discourse markers, Gibson's speech is not greatly constrained by the interview situation.

The following extract gives an example of Gibson's more fluent style in narrative:

(14)	WG1102	and there is such a thing as rough speech
	1103	I—I remember oh about fifteen years ago
	1104	I—I became friendly with—a chap
	1105	who's a miner at New Cumnock
	1106	and in the event—in—the day came
	1107	when he said
	1108	"Oh—ah—come up
	1109	spend the weekend"
	1110	which I did
	1111	I found it quite traumatic
	1112	talk about class
	1113	there is such a thing as class
	1114	I—I always think
	1115	it's very foolish
	1116	to say
	1117	there isn't class
	1118	whether you make class—ah—the distinctions of class rule your life
	1119	's a different thing
	1120	but it's nonsense
	1121	to say
	1122	there isn't class
		[23 lines omitted]

1145   but going out to New Cumnock
1146   it was a completely different—environment [low
       breathy voice]
1147   and the way
1148   they talked
1149   they tended
1150   to shout at each other
1151   "Aw Jeannie—"
1152   and they also had the TV on all the time
1153   I had
1154   to shout above the TV
1155   nobody seemed
1156   to communicate with anybody—except in these
       basic—
1157   one of the things they did
1158   as I later discovered
1159   [imitating] "You know Stanley
1160   ah've got a feeling
1161   Big John will come and see us the day"
1162   and he had
1163   to motor from somewhere in mid-Scotland [draws
       breath in sharply]
1164   and they would—keep
1165   looking out the window
1166   and saying
1167   "Ah've got a—" you know
1168   on subsequent meetings—visits this continued
1169   "Ah've got a feeling Big John—" you know
1170   Big John never appeared
1171   then midsummer I was there
1172   when lo and behold Big John did appear
1173   "We were just saying to Stanley
1174   ah've got a feeling
1175   Big John'll come"
1176   and I thought
1177   "Aye and you've been feeling it since December"
1178   but—these—I—
1179   coming down on the bus
1180   I can remember this
1181   I'm smiling to myself
1182   the Black Bull in Ochiltree
1183   I read

1184   as I passed in the bus
1185   I read it
1186   and heard it in my head as the [blɑk bʌl]
1187   because I'd been exposed to this kind of talk all
        weekend
1188   I can remember a curate's wife
1189   coming up in the car
1190   to join us there
1191   and go out on a picnic
1192   and the whole family and a kid brother
1193   that had come in
1194   struck dumb
1195   because she was English
1196   [imitating] "Oh hello how are—"
1197   absolutely mute goggling at this
1198   so there was a complete difference there

This passage is a good example of Gibson's style. There are eight dysfluencies in the first 21 lines but only six in the next 56 lines (three of them in line 1178 at the transition from the main subject to the subordinate example). Essentially, when Gibson gets into the flow of the narrative, the jerkiness vanishes and he is in complete control of his vocal tract. The truncated phrases in quoted direct speech are not dysfluencies but abbreviated forms to avoid redundancies. Unlike some of the lower-class speakers, Gibson does not feel it is necessary to repeat every syllable of a quoted dialogue. This is indicated by his use of *you know* in lines 1167 and 1169. As shown in Table 10–5, there are only seven examples of *you know* in Gibson's interview, two of which occur in this passage. Their use is a good example of Östman's notion of camaraderie, since in effect Gibson is saying that he doesn't need to repeat in detail what was said because he knows that I can fill in the missing elements for myself. The passage is relatively complex syntactically, with 35% subordinate clauses and 14 nonfinite structures. There are also 13 lines of quoted direct speech.

The evidence suggests that this interview is a reasonable sample of Gibson's speech in one of the styles that he claims to use. Speaking to members of the community that he would place in a different category from the interviewer, his speech would be somewhat, but probably not very, different from what was recorded here. This is a very satisfactory sociolinguistic interview. If all the interviews in sociolinguistic surveys were of this quality, it would be safe to make generalizations with the confidence that they had some basis in reality.

The Nicoll, Gemmill, Laidlaw, Lang, and Sinclair interviews, although they are each very different, can be considered unqualified successes in that they provide extensive evidence of the speakers' use of language. It is important for the claims I am making that these interviews include speakers from very different backgrounds and at least one woman. They provide evidence that the range of speech that can be collected through interviews is sufficiently impressive to justify the use of such an approach in sociolinguistic surveys, despite the criticisms of Wolfson and the Milroys. The first to be considered is the most extensive interview from the middle-class group.

Nicoll

18 topics initiated by interviewer = 22%
63 topics initiated by respondent = 78%
Proportion of interview initiated by respondent = 82%
Average length of topic = 35 syntactic units
Number of narratives = 15
Length = 15,268 words

Compared with the other middle-class respondents, Nicoll was a joy to interview. In fact, it would be more accurate to say that he talked to me rather than that I interviewed him. His interview is the longest in the middle-class sample, yet I had the opportunity to ask only 11 questions. About a third of the way through the session, we moved from his place of work to his house, a few doors down the road. As he introduced me to his wife he said, "I haven't found out what he's doing it for"; and the reason was that he had barely let me get a word in up to that point. As MacGregor remarked of Nicoll, "He is one Scot who isn't reticent"; and at one point Nicoll himself remarked, "I could go on for ever." In the first fifth of his interview there are only four turns, averaging 160 lines each. More than a quarter (26%) of his turns are more than 30 lines long, eight are more than 100 lines long, and his longest turn is 379 lines long. Unlike most of the speakers, Nicoll is expansive from the very start and his average length of turn drops most noticeably when we move to his house and there is a brief exchange with his wife over coffee. Like the Belfast speakers reported by the Milroys, Nicoll takes up a narrative again exactly at the point where it was interrupted by other discourse.

Nicoll is the only member of the middle-class group who did not finish secondary school; he left at the age of 14 or 15, by his own choice, and went into his father's tailoring business. He had been in a variety of occupations in which an ability to talk would be an advantage, and he clearly enjoyed the opportunity to talk at length.

As has been shown in several tables in earlier chapters, Nicoll's language in the interview is the closest of all the middle-class speakers to that of the lower-class group. He has the highest percentage of coordinate clauses in the middle-class group and the lowest proportion of subordinate clauses; he has the lowest percentage of subordinate noun clauses; and his use of subordinate adverbial clauses does not show the pattern found in the other middle-class speakers. On the other hand, he has the typical middle-class frequency of infinitives and the second highest proportion of nonrestrictive relative clauses. His use of discourse markers matches the average for the middle-class group. He has the second-highest percentage of glottal stops in the middle-class interviews; but segmentally, his speech is very similar to that of the other middle-class speakers. As pointed out in Chapter 7, the differences between the language of Nicoll's interview and that of the other middle-class speakers is probably largely the result of genre. The proportion of narrative in Nicoll's interview (45%) is almost twice as much as that in the two other middle-class interviews with a substantial amount of narrative (MacGregor's and Gibson's). Nicoll also has the highest proportion of quoted direct speech (16.2%) of the whole sample, including the lower-class speakers. Except for his love of talking (a trait not widely manifested in the other middle-class interviews) and his skill in storytelling, Nicoll is not very different from the other middle-class speakers in the sample.

The richness of the interview makes it harder to choose a representative sample of Nicoll's speech, but I have chosen a relatively short one because it ends with a moral that is very dear to Nicoll's heart. I plan to deal with Nicoll's narrative skills (and those of other effective storytellers) at greater length in a subsequent work.

(15)   DN1804   my father uh was a very—
         1805   my father loved
         1806   to tell the story
         1807   that he started his work in his bare feet
         1808   he was a barefooted boy
         1809   and there was a J. C. Hyatt
         1810   who had a plumber's business down the same close a
                 wide close
         1811   which my father went down
         1812   to start his sewing at six o'clock in the morning
         1814   and this boss of the plumber's business was a gen-
                 tleman a well-to-do business man
         1815   but he always had time at ten-to-six
         1816   to say good morning to my father
         1817   who was a barefooted boy
         1818   he says

1819   "You're late this morning young man
1820   it's nine minutes to six
1821   you were here at ten minutes to six yesterday
1822   nice to see you"
1823   and my m—father never forgot that
1824   the best people could always speak to the ordinary
        people
1825   he was just a barefooted boy
1826   and he could speak to the ordinary people
1827   and it's so nice
1828   to be able
1829   to speak to anybody

Despite the fact that Nicoll messes up the punch line (he presumably means in lines 1825/26, "He was just a barefooted boy and he could speak to the *best* people"), this is a good example of Nicoll's narrative style—fluent and economical, with some quoted direct speech, which he enunciates in appropriately formal tones. But the conclusion stated in lines 1827/29, "It's so nice to be able to speak to anybody," is the theme that runs through the whole interview.

Nicoll's interview is particularly important for this microsociolinguistic study of Ayr speech because without it, the narrative ability of some of the lower-class speakers would have appeared to be even more strongly class-linked. In the long run, the lower-class narratives are more interesting, more amusing, and better-told; but Nicoll and, in their own ways, Gibson and MacGregor show that middle-class speakers can be effective storytellers, too. The most effective lower-class speakers, however, show that narrative skills are widespread.

Gemmill

19 topics initiated by interviewer = 45%
23 topics initiated by respondent = 55%
Proportion of interview initiated by respondent = 67%
Average length of topic = 42 syntactic units
Number of narratives = 10
Length = 9,761 words

Gemmill is the oldest speaker in the group and, as a farm laborer, probably the one who is most similar to Wilson's "authorities"—though Gemmill's later experience had been very different from anything they could ever have known. After explaining to me that he didn't get vacations as a young man, he told me about how he goes to Spain now. The thought of someone who had grown up before there was even a local bus service getting on a plane to fly to Spain intrigued me, and I asked if he had felt frightened. He told me about one occasion when there was thick cloud, and he wondered how the pilot was going to land; but all went well, and he concluded:

(16) HG1201 well it didnae frichten me in any case
1202 the other point is
1203 that folk get frichted
1204 they'll get killt
1205 heh
1206 you can faw ten feet and get killt
1207 you don't need to faw ten thoosand feet
1208 for to get killt

There were many other similar comments about life, past and present, that sounded like the kind of thing Wilson might have heard from his respondents.

In phonology and morphology Gemmill is the closest of the lower-class group to the variety recorded by Wilson. In his use of the variables (a), (e), and (au), the velar fricative /x/, and the vocalization of /l/, in addition to final devoicing of −ed and the loss of homorganic voiced stops after nasals, Gemmill is very similar to Wilson's speakers, although he must have been born at least 50 years later. Gemmill also resembles them in his use of nae, no, −nae, yin, and thae, as well as in some of the verb forms.

Gemmill's interview consists mainly of recollections of what it was like to be a farm laborer in the early part of the twentieth century; but there are also 10 narratives, making up approximately 15% of the transcript. His turns tend to be long (average, 29.7 lines), with five turns over 100-lines long, including one of 166 lines. He has the highest proportion of coordinate clauses, the second-highest percentage of subordinate clauses, and the lowest percentage of that-deletion in the lower-class group. Gemmill's language does not noticeably vary throughout the interview in phonology, morphology, or syntax. An example of his narrative style is shown in (17):

(17) HG661 I had another experience of this
662 they had grazing doon at Bargennie
663 and this day
664 it was an Irishman
665 that was working aboot the place
666 and the two o us was to go to Dailly
667 to pick up half a dozen horse and a dozen bullocks
668 and we went up to Dalrymple

669   to get the train
670   and we got the train
671   and I had aye heard tell o the burning pit
672   so I must see this burning pit
673   I seen the burning pit aw right
674   I got a spark in my eye
675   and anyway I stertit—
676   the worst thing I could have done
677   I sterted rubbing it
678   so anyway we got the horse and the beasts
679   and set oot
680   we werenae too bad
681   until we came to Meybole
682   and when we come doon to the bottom of Meybole
683   the blooming things took away doon for Crosshill
684   we couldnae leave the horse
685   anyway we got them awthegither again
686   and set oot
687   and we were going to Culroy to the smiddy there wi
       some of the horses
688   and there were some o them going into a field
689   so anyway we went ower the brig at Blackhill
690   and there a gate
691   just efter you go ower the brig
692   takes you intae this field
693   but here the roadmen were working on the side of the
       road
694   by this time I had my bunnet drew doon ower my eye
695   and this auld yin says
696   "Whit's wrang wi your face?"
697   I telt him
698   he says
699   "Let me see it"
700   so he says
701   "Get doon there on the grass"
702   so I got doon on the grass
703   and he gets oot this knife
704   and he took it oot
705   but the unfortunate thing was
706   he'd been cutting tobacco wi it
707   before he did this

708   it was sair before
709   but it was a dashed sight sairer efter
710   but in the long run it sortit it oot right enough

The phonological and morphological variables are consistently lower-class. Note also the plural *horse* in lines 667, 678, and 684, compared with *some of the horses* in line 687. In line 678 the *beasts* refers to the bullocks and is a common word for members of the bovine species. Gemmill also illustrates the use of two of his favorite discourse markers: *anyway* in line 685 and *here* in line 693.

There are few signs anywhere in the interview that the situation or the presence of the tape recorder or the speech of the interviewer is affecting Gemmill's language. Thus, the fact that there is still variation in his use of certain features is probably not an artifact of the interview. Probably, Wilson's authorities would have shown similar variation if he had been looking for it. Accordingly, it seems reasonable to assume that the evidence in the interview is a typical sample of Gemmill's speech.

When at the very end of the interview I asked if he had ever thought of moving away from Ayr when he was younger, he replied:

(18)   HG1753   naw I never really got that thought at aw
        1754   naw mair inclined
        1755   to work away in the auld tradition
        1756   if you can caw it that

The similarity of Gemmill's speech in the interview to that of Wilson's "authorities" suggests that in language, also, Gemmill was "in the auld tradition."

Laidlaw

9 topics initiated by interviewer = 15%
52 topics initiated by respondent = 85%
Proportion of interview initiated by respondent = 68%
Average length of topic = 45 syntactic units
Number of narratives = 14
Length = 13,189 words

Ella Laidlaw made it clear at several places in the interview that she was aware of variation in language and how important it is to be able to speak appropriately. At one point she remarked, referring to a meeting I had attended but, to my chagrin, was unable to record, "You saw how we can all sit and talk wer ain tongue wer ain way among werselves." And she referred to herself as a "broad Scotch" speaker; but she pointed out that if you were talking with "the cooncillors you've got to keep your—match them sort of style but when I'm among my ain I just talk whatever wey that suits me." Laidlaw had also been in domestic service with a middle-class family and probably did not speak her "ain way" with them. Laidlaw had been a promising pupil at school, and her mother had wanted her to stay on; but Laidlaw had resisted the idea. One of her stories refers to a boy not more successful than she at primary school, who later became a professor at Glasgow University. Since she is intelligent and aware of the significance of language differences, it is not surprising that her speech during the interview should vary. As was pointed out in Chapter 4, Laidlaw's use of a monophthong for the variable (au) increased in the second half of the interview and even more strikingly in a narrative section. There are 14 narratives in Laidlaw's interview, making up approximately 30% of the whole; and she has the highest proportion of quoted direct speech (11.5%) in the lower-class interviews. She has six turns of more than 100 lines, the longest being 398-lines long, the longest single turn in the whole set of interviews. The following narrative about her mother comes late in the interview:

(19)	EL3633	I used to be petrified
	3634	to go in five-minutes late
	3635	because she was behin the door
	3636	waiting on you—wham!
	3637	I remember
	3638	she always warned me
	3639	I wasnae to get myself into ony bother
	3640	just because of hersel
	3641	and I said
	3642	"All right"
	3643	but she never trusted you—never trusted you
	3644	she watched you like a hawk
	3645	so I goes oot this night
	3646	it was my first husband
	3647	I'd made arrangements
	3648	to meet him—away at Tam's Brig
	3649	away from the Prestwick Road to the Tam's Brig
	3650	and somebody had telt her
	3651	they had seen me
	3652	so we'd made arrangements
	3653	we'd meet at Tam's Brig
	3654	he would go his road

3655   and I would go mine
3656   and then naebody would see us
3657   walking hame
3658   however whoever spouted on me
3659   had telt her
3660   where I was and aw the rest of it
3661   so she—
3662   I come to Tam's Brig this night
3663   and I'm just coming ower Tam's Brig
3664   and I stopped dead
3665   Bertie says to me
3666   "What's up with you?"
3667   I says
3668   "Oh don't luck the noo
3669   there's my mother"
3670   he says
3671   "It is nut"
3672   I says
3673   "It is"
3674   "Well come on
3675   and we'll face her"
3676   I says
3677   "You may
3678   but I won't"
3679   says I
3680   "You better stop there
3681   and I'll go on"
3682   he stopped
3683   as I telt him
3684   and I went on
3685   well she hammered me fae the Tam's Brig tae the Prestwick Road
3686   and everybody watching me
3688   and I was eighteen
3689   so you ken what
3690   that's just
3691   how we were brought up
3692   if you didnae respect it
3693   you respected the strap
3694   that come oot or the stick
3695   whatever came first

This is Laidlaw at her most "Scotch" in the interview, consistent in her use of /u/ for (au) and /e/ for (e), the vocalization of /l/, and the negative clitic *–nae*. Note also /ʌ/ in *look* (3668), and the forms *telt, fae, ower, ken, ony,* and *behin*. She also uses the "narrative present tense" in line 3645, "So I goes oot this night."

Compare the language in Item 19 with that in Item 20, which comes early in the interview and before her husband had entered the room:

(20)   EL349   we had a picture house maybe once a week
         350   but very very strict as far as the church was concerned
         351   I was brought up
         352   I was made
         353   go to the Sunday school
         354   whether I wanted
         355   to go or not
         356   when I came out of there
         357   I wasn't allowed
         358   to play games on a Sunday
         359   we went—
         360   my mother maybe or some of them took us for a walk
         361   and we went back to the evening class at night
         362   what we called the Christian Endeavour
                 (What happened there?)
         363   just a Bible reading—a hymn—
         364   the minister or someone in—well up in the church
                 would give a wee bit of a lecture
         365   you were dismissed
         366   and Heaven help you
         367   if you didnae go straight home

Here (au) is a diphthong in *house, out,* and *allowed; home* has /o/; there is no vocalization of /l/ in *called;* the negative clitic is *–n't* in line 357 (but *–nae* in line 367); the isolated negative in line 355 is *not;* and the past tense of *come* is *came.*

The differences between Items 19 and 20 are clear, but it is important not to exaggerate their salience. It is not as if Laidlaw had "switched" from one "code" to another. It is impossible to tell from the language alone at what point the change occurs. Most of the segmental features and prosodic features remain the same and in a "matched guise" test there would be little doubt that the same person was speaking. The differences between the two extracts are more like such stylistic differences as cliticization of auxiliaries and negatives. Variation can occur within the same context. Almost at the end of the interview occur the following lines:

(21)   EL3869   they're no doing theirsels any good
         3870   but they're daeing it all over the world

Here in line 3869 *doing* cooccurs with *no,* but also with *any* and *good.* In the next line *daeing* cooccurs with *all* and *over.* This is not part of a narrative but a section in which Laidlaw is offering an opinion on the state of the universe. Such variation is possible because the alternatives exist within the same framework. It is not necessary to "code-switch" in order to make the choice.

Given that Laidlaw's interview shows greater variation than Gemmill's, it is necessary to treat the evidence somewhat differently. In some cases, where alternative forms are fairly frequent, the percentage of use may be of interest. In other cases, where the forms are relatively infrequent, the significant point is the occurrence of a particular form, since that indicates that it is part of Laidlaw's language and is available as an alternative.

Laidlaw's use of the three phonological variables that were quantified in Chapter 4 shows a similar pattern: (au) 48%, (e) 41%, vocalization of /l/ 47%, and (gs) 35%. In negation she has 8% *–n't* (vs. *–nae*), 38% *not* (vs. *no*), and 86% *no* (vs. *nae*). All these figures show some variation away from the pattern of Gemmill's speech (and the account given by Wilson) in the direction of middle-class norms.

In the (a) variable Laidlaw has examples of *aipple, caird, Chairlie, cless, efter, gairdens, gled, hairm,* and *maiter,* which displays good evidence of the basic system. She has *guid* and *shin* (= *soon*) but no other examples of /ɪ/ for (o). She has *dee, gie'd, heid, neebor,* and *weel; pey* and *wey; luck* (= *look*); *wuman;* and *wallop* and *want* with /a/.

What Laidlaw does not show is any example of the velar fricative /x/ or the loss of homorganic voiced stops after nasals (except for one example in a proper name); and there is only occasional devoicing of *–ed.*

Laidlaw's interview provides a good sample of her speech, and the presence and interventions of her husband no doubt helped to counteract the effect of the interview situation. Nine years younger than Gemmill, with a very different kind of upbringing and work experience, Laidlaw nevertheless displays a very similar basic language system; but she also has a wider repertory of choice for stylistic or other purposes.

Lang

21 topics initiated by interviewer = 23%
72 topics initiated by respondent = 77%

Proportion of interview initiated by respondent = 82%
Average length of topic = 32 syntactic units
Number of narratives = 24
Length = 16,255 words

With 24 narratives making up about half of the transcript of an interview of 16,255 words, this interview was a pleasure at the time and has continued to be one of the most rewarding throughout the often-tiresome process of analysis. Lang needed little prompting; almost a quarter (24%) of his turns are 50 lines or longer, with the longest 241 lines. As would be expected from the large number of narratives, Lang has the highest proportion of coordinate clauses (31.6%) and the second-lowest proportion of subordinate clauses (18.4%). Like Gemmill, Lang shows little variation throughout the interview; but since he is nearer to me in age, there was less distance between us—and his narrative style reflects this (for example, in his very frequent use of *you see*). Lang has a wider range of narratives, partly because he tells more stories but also because that is his usual way of dealing with a topic. He is very similar to Gemmill in his use of (au), (e), and (a); and their use of glottal stops and the velar fricative /x/ is much the same. Since Lang is 22 years younger than Gemmill, the similarity between the two is striking. Such differences as are observable are probably more due to differences of topic and genre than to linguistic changes. An example of Lang's narrative skill, including his use of quoted direct speech, is given in (22):

(22)   WL654   in fact there were a fellow doon there
       655   Andy Smith you caw'd him
       656   he belonged to Girvan
       657   my neeghbor come
       658   and he says to me
       659   "Willie, you'll need to come up to Curly"
       660   Curly—was it Curly you caw'd him?
       661   aye Andra Smith
       662   it was Curly
       663   they caw'd him
       664   his nickname
       665   he says
       666   "Cannae see him
       667   I think he's buried [t]"
       668   so away I goes
       669   running
       670   and this big stane was doon you know

671   I says
672   "Where is he?"
673   "Oh" he says
674   "he's in there"
675   I says
676   "Oh naw"
677   "Aye he's in there"
678   so up ower the tap of this
679   and lucking up
680   and watching
681   and he's lying away in the inside
682   he heard it
683   coming
684   and if you hear something
685   going to fall
686   automatically you run oot the way
687   but he didnae
688   he went in the wey
689   see if he'd tried
690   to went oot
691   that wouldn've come doon and got him
692   he went in the way
693   and he was sitting in a space like that
694   but it had knocked the battery off his lamp off his
        hinge on his belt
695   and the belt was smashed sort of thing you see
696   he was lying in here
697   so I says to him
698   "Are you aw right?"
699   "Aye" he says
700   "I'm no bad"
701   says I
702   "Well I don't ken hoo"
703   says I
704   "Put your haund roon my neck
705   and I'll pu ye up"
706   in this narrow space
707   so I lifted him ower this
708   and I shouted
709   to get a streetcher and splints you see

710   of course I didnae ken
711   what was wrong with him
712   but you needed a streetcher
713   I kent
714   he was needing
715   to be cairried [t] you see
716   so when I got him ower here
717   and we put the streetcher doon
718   and fower of us were going to cairry him oot
719   "Right boys we'll coont three
720   and then we'll lift aw thegether"
721   "One two three"
722   and we lifted
723   so we lifted the streetcher
724   and he was left
725   lying on the pavement
726   it was rotten you see
727   the streetcher was rotten
728   well after that they brought oot thae kind of wire net
        things
729   that was steel
730   that was only a widden yin with a cloth on it you see
731   oh we just lifted him
732   he was that heavy
733   we just lifted the streetcher
734   and he was left
735   lying there
736   so we had to make do with jaickets and one thing and
        another
737   wer jaickets off
738   and tied them on
739   to get him cairried to the pitheid aye

This passage shows that in almost every instance of the key variables Lang is consistent with the evidence from Gemmill's interview and with the account in Wilson. The exceptions are *wrong* (711) and *off* (698, 737; cf. *tap*, 678), *after* (738), and *where* (672)—though elsewhere Lang uses the older forms cited by Wilson.

Lang's interview confirms the evidence of Gemmill's and suggests that relatively few changes in the basic phonological system have taken place since Wilson's study.

Sinclair

34 topics initiated by interviewer = 44%
43 topics initiated by respondent = 56%
Proportion of interview initiated by respondent = 50%
Average length of topic = 24 syntactic units
Number of narratives = 13
Length = 21,163 words

At 21,163 words and lasting two hours and twenty minutes, Sinclair's interview is the longest in the whole sample. He has by far the longest average turns (60.2 lines); almost half his turns (43%) are 50 lines or longer, with the longest being 300 lines. With 13 narratives making up approximately a quarter of the whole, the interview also provides Sinclair with plenty of opportunities to display the full range of his language; but it is not clear that the spectrum is complete. When the arrangements for the interview were being made he commented:

(23)	AS001	I don't think
	002	I would sound as broad to you
	003	as I would to other people
	004	this is where
	005	after being in the likes of community associations et cetera and that for quite a number of years
	006	eh—you begin
	007	to—well is the right word *learn?*
	008	you begin
	009	to realize
	010	that it just doesnae do
	011	for to speak in your normal way
	012	that you would to everybody else
	013	you've got
	014	to try and compromise
	015	as I would say and all like that

There is an ellipsis in the first three lines. Sinclair is saying that he probably will not speak as broadly talking to me as he would talking to other people. This is similar to Laidlaw's statement quoted above; but Sinclair applies his remarks directly to the interview situation, and his behavior in the interview bears out the accuracy of his prediction. It is easier to see the effect on Sinclair's speech because of the direct comparison with Lang and Rae. All three had been coal miners and were roughly the same age. Sinclair, however, had spent two years in his late teens away in the merchant navy during the

Table 12.6. Lang, Rae, and Sinclair:
Proportion of Selected Features (%)

Feature	Lang	Rae	Sinclair
(au)	87	83	31
(e)	81	91	20
(x)	38	14	2
(gs)	66	84	28
voc. of /1/	66	91	37
no (vs. *not*)	85	93	36
nae (vs. *no*)	52	76	26
yin/wan	27	62	5
subord. cl.	18	16	36
nonagreement s/v	9	13	6

war. Lang and Rae had never spent any significant period of time away from Ayr. Some of the differences can be seen in Table 12–6. There are other differences, but those shown in Table 12–6 are enough to make the point. In terms of the overall structure of the interviews, Sinclair's and Lang's are much more similar to each other as regards length and variety than either is to Rae's; but in linguistic terms Lang's and Rae's are extremely close and very different from Sinclair's. The linguistic differences are not obviously linked to topic or genre but are more likely to be the consequence of Sinclair's accommodation to a middle-class interviewer, as he had warned me. The interesting point is that he sounds perfectly relaxed and natural and not very different from the way he spoke to his brother on the telephone during an interruption of the interview. To outsiders, such as my students, Sinclair does not sound noticeably different from Lang or Rae. Like Laidlaw, Sinclair manages to accommodate to the middle-class interviewer without making it at all obvious. I have discussed this interview at length elsewhere (Macaulay 1985). The extract in (24) gives a good illustration of Sinclair's style:

```
(24)            (You were talking about the merchant navy)
        AS2271  well how I went into the merchant navy was that
          2272  as I told you already
          2273  I mean I was doing a man's work
          2274  fae I was fourteen years of age
          2275  I mean as a boy of sixteen I was on night shift
          2276  constant night shift aw the time
          2277  I mean you'd
          2277  to run
```

2278 it doesnae matter
2279 what you were doing
2280 you'd
2281 to run away at nine o'clock
2282 to get home
2283 to change yoursel and everything
2284 to go to your pit
2285 you got your bus at ten o'clock
2286 to go to the pit
2287 ach I got fed up and that
2288 and I went away to Glesca
2289 I steyed there for two or three weeks
2290 I was going to sea you know
2291 well I didnae know the setup for that
2292 and you went
2293 to see aboot
2294 getting jobs on boats and all these things
2295 and I was told
2296 "Oh you've to jine the union"
2297 I went away
2298 and I joined the Seamen's Union
2299 and you used to go to
2300 what they called the pool place
2301 where aw the people were signed on for ships and
      that
2302 and och you got turned down
2303 it was a case of
2304 not what you know
2305 who you knew kind of style
2306 but I happened
2307 to be there one day
2308 and they were signing on a crew for a ship
2309 it wasn't in this country
2310 well as you know
2311 things was getting bad just the first year of the war
2312 there was so much shipping being sold
2313 and they were bringing everything
2314 that they could
2315 back to sea and that
2316 and this ship it was lying in Philadelphia in America
2317 the *S. S. George Washington* they called it

2318 and she was a—actually a German liner
2319 that had been captured in the First World War
2320 and America had used her
2321 and she'd been tied up
2322 and hadnae done nothing for years
2323 and this was them
2324 Britain was getting
2325 to put back to sea
2326 and they were signing on a crew
2327 to go
2328 and bring this back to Britain and one thing and another
2329 and as it was a coal-burning ship
2330 I mean a liner of twenty-six thoosand ton
2331 but she still burned coal and that you see
2332 and they were lucking for stokehole crew
2333 and they couldnae get them
2334 of coorse when I came up with my union caird and that
2335 he looked at me and that
2336 and he says
2337 "What did you dae before?"
2338 "Oh" I says
2339 "I used to be a miner"
2340 he says
2341 "Let me see your haunds"
2342 and he lucked at my haunds
2343 and he says
2344 "Right you are" he says
2345 "You'll dae"
2346 so I was signed on as a stoker you see
       (You can't get away from coal!)
2347 that was right you see

This gives a good idea of the variation that occurs in a short section of Sinclair's interview:

(25)	how 2271, down 2302	aboot 2293, thoosand 2330, coorse 2334
	doing 2273, 2279	dae 2337
	joined 2298	jine 2296
	matter 2278	caird 2334, haunds 2341/2

called 2300, 2317;	aw 2276, 2301
all 2317	
looked 2335	lucked 2342, lucking 2332
change 2283	steyed 2289
home 2282	

Some of this variation could be deliberate, since some forms occur only in quoted direct speech (e.g., *jine* 2296, *joined* 2298); but that will not account for all the variation, and (as was pointed out in Chapter 11) sometimes Sinclair's use of forms in quoted direct speech is difficult to explain. The more likely explanation is Sinclair's own, that he is not "as broad" in talking to me as in talking to others; but in narratives, quoted direct speech, and certain kinds of asides this constraint seems to be weaker, though never totally absent.

Sinclair is not only a voluble talker, he is truly long-winded. Not only does he go into great detail in his explanations, he also is not afraid of redundancy in his actual language. This gives a characteristic style to many of his utterances. Some illustrative examples are given in Item 26:

(26) a. AS1033 I *never ever* was afraid
      b.   813 it's *mostly more or less* liquid food
      c.   1011 it didn't pay you to be *too confident* or *overconfident*
      d.   1124 you'd only *roughly* oxygen to last you for *roughly* two hours
      e.   1303 you *possibly maybe* had to leave
      f.   1407 in that case there were *usually possibly* two of youse worked thegither
      g.   2101 you maybe had twenty-four dozen bottles of beer *et cetera and aw the rest and that there*
      h.   3173 but there *many a time* my mother *often* said

Like many of the other features of Sinclair's speech, this kind of repetition gives his language a slightly formal, or "academic" quality. While the interview situation may have caused Sinclair to emphasize this quality, it is highly unlikely that the speech event alone is responsible for this repetitive style. Similarly, the higher proportion of subordination and other syntactic features that bring the language of Sinclair's interview closer to that of the middle-class group are probably not the result of accomodation to the interviewer or the interview situation but reflect more characteristic aspects of his language. Sinclair is the working-class speaker who says most about such relatively abstract topics as education, politics, and labor relations. Most of his remarks on these topics were self-initiated, and many of them sounded as if this was not the first time he had made them.

Sinclair's interview shows the greatest internal variation of all the lower-class group. As far as the phonological and morphological variables are concerned, he is not as "broad" as he would be talking to his friends; but the other aspects of his speech may be less affected by the situation. Sinclair's control of the phonological and morphological variables is presumably the result of his wider experience of the world. As pointed out in Chapter 2, Sinclair had spent two years in the merchant navy in his late teens. During that time he would no doubt have been obliged to modify his speech more consistently and for a longer time than any of the other lower-class speakers. It is therefore not surprising that in contact with middle-class speakers (as in the community associations he refers to or in my interview with him) he should be able to modify his speech in a "convergent" direction (Giles and Smith 1979). At the same time, he shows, for example, in the use of the variable (a), that he has not abandoned his earlier form of speech.

As evidence on which to build a description of lower-class speech in Ayr Sinclair's interview, like those of Laidlaw and Ritchie, should not be taken alone but conjoined with the interviews of Gemmill, Lang, and Rae. These latter three provide a wider perspective of the range of lower-class speech. In particular, they show how easily variant forms can be accommodated within a single speech event.

I have looked at the individual interviews in an attempt to assess their value as evidence for the dialect of Ayr. While it is obvious that some of the speakers were more at ease in the interview situation than others and that this affected both the quantity and quality of the speech recorded, there seems no reason to reject any of the interviews as illegitimate pieces of evidence. Taken together, the 12 interviews present a consistent picture of the dialect, showing the range of linguistic choices available within the community.

In addition, a close examination of the individual interviews refutes the views that the speakers are generally at a loss in the interview situation (Wolfson 1976), that "interview speech" consists solely of short answers to questions (Milroy and Milroy 1977), and that the interviewer is in constant control of the speech event. In his account of interviews in social science, Briggs claims: "Control over the interaction lies in the hands of the interviewer. It is he or she who exerts the most control over the process of turn-taking, it is he or she who introduces topics, by and large, and it is he or she who decides when to move onto the next topic" (1986, 89). As the preceding account has shown, this is true only of some of the Ayr interviews. Briggs, of course, was describing interviews in which the elicitation of information is the primary motivation, particularly those in which the interviewer is not completely at home in the culture (or even the language) of the community. Briggs's recommendation that the investigator needs to understand the norms of speech in the community is very important; but given the basic understanding of the situation that anyone undertaking a sociolinguistic must have, there is no reason to reject the valuable evidence that interviews can provide.

13

The Dialect of Ayr

The preceding chapters have presented some of the evidence of Ayr speech that can be gleaned from the 12 interviews. Varied as this information is, it by no means exhausts the storehouse that the tapes provide—the analysis could continue to probe deeper and deeper into the material; but it is reasonable at this point to try to take stock and decide what has been accomplished by the kind of analysis presented here. In particular, it is important to decide whether it is legitimate to consider the speakers as representing in any sense a coherent linguistic group rather than simply a collection of disparate individuals. The first point to be made is that the individuals fall into two groups, clearly distinguished by their speech. The linguistic analysis shows that despite intragroup variation, the two groups show differences in their speech that support the allocation of the speakers to one or the other group. The social class differences found in the interviews are summarized in Table 13–1.

The items in sections I and II are relatively "robust," in that they distinguish the speech of the two groups regardless of topic, style, or genre. The items in sections III and IV are much more "fragile," since their frequency depends much more on topic and genre. In fact, Table 13–1 is arranged roughly in order from most robust to most fragile. It summarizes the main features that vary with social class membership. The pattern is unaffected by individual deviations from the general pattern such as Sinclair's use of WH–relative markers or Nicoll's occasional use of the clitic –*nae,* since some intragroup variation is to be expected in social class groups (Macaulay 1978a). This volume, however, is an attempt to provide a microsociolinguistic study of the dialect of Ayr. Given the considerable differences that exist among the speakers, to what extent is it legitimate to claim that all 12 respondents speak the *dialect* of Ayr?

The term *dialect* has proved to be problematic for modern linguists, so

Table 13.1. Summary of Social Class Differences
in the Ayr Interviews

Feature	Lower-class	Middle-class
I. Monophthongal (au)	very common	absent
Variable (e)	common	absent
Variable (a)	common	absent
−ed devoicing	common	absent
Homorganic stop deletion	common	absent
Vocalization of /l/	common	very rare
Glottal stops	more	fewer
Velar fricatives	more	fewer
II. Clitic −nae	very common	absent
neg. operator no	very common	absent
Subj./verb nonagreement	common	absent
Multiple neg. concord	rare	absent
III. WH−relative markers	very rare	very common
Nonrestrictive relative clauses	rare	common
Subordinate clauses	fewer	more
Noun clauses	fewer	more
Subordinate clauses: Reason, condition, concession, embedded questions	fewer	more
Comparison, place, time	more	fewer
Infinitives	fewer	more
Gerunds	slightly fewer	slightly more
IV. Highlighting devices	more	fewer
Adverbs	fewer	more
Discourse markers	more	fewer

much so that some have even despaired about its value. Bickerton, for example, puts the point fairly strongly when he claims, "To speak of 'dialect' or even perhaps 'languages' may be misleading; these terms merely seek to freeze at an arbitrary moment, and to coalesce into an arbitrary whole, phenomena which in nature are ongoing and heterogeneous" (1973, 643). Bickerton's experience with pidgin and creole languages may have led him to exaggerate the heterogeneity of language; but Hudson, from a rather different perspective, takes an equally extreme position: "We can be sure that *no two speakers have the same language, because no two speakers have the same experience of language*" (1980, 12, emphasis original). In a trivial sense this is true of all human behavior and individual characteristics, but that does not mean that all generalizations are pointless. Hudson himself, even though he believes that "constructs such as 'language' and 'dialect' have little or no objective reality" (p. 232), admits that there are perplexing questions: "Why do people in the same community often speak in such a similar way, down to the most minute details of pronunciation? How do they achieve such precise agreement? Why do they differ with regard to some particular variables?" (p. 233).

Unfortunately, the notion *speech community* is equally problematic (Labov 1966a; Gumperz 1968; Romaine 1982b; Petyt 1982). Labov claimed that New York was "a single speech community, united by a common set of evaluative norms" (1966a, 500). Trudgill in his study of Norwich speech—believing that the speakers belong "to a particular speech community which has very many common linguistic features," many of which "are not shared by any other speech community"—attempts "to establish a common framework that will, in some sense, incorporate all types of Norwich English. We hope to show that these different types are derived from a single underlying framework that is the same for the whole community" (1974, 133–34). Trudgill's position was partly based on the ability of Norwich speakers to imitate those who spoke a different variety of Norwich speech. More recently, Trudgill has argued against panlectal or polylectal grammars on the grounds that "it is a rare speaker who acquires or imitates a dialect other than his own vernacular perfectly" (1982, 189). In this statement Trudgill is apparently using *vernacular* in the sense of idiolect or mother tongue, as suggested by Labov: "Each speaker has a vernacular form, in at least one language" (1981, 3). Another use of *vernacular* is in the sense of *basilect,* for example, "Belfast vernacular . . . is an example of a highly focused variety at the lowest stratum" (L. Milroy 1980, 180). It turns out, however, that Milroy's characterization of Belfast vernacular is so narrow that none of her speakers actually uses it consistently. Similarly, Heath (1980) found no speakers of "pure" Cannock dialect in his study.

In a similar way, it would be possible, based on several of the lower-class interviews, to specify a form of Ayr speech that none of them uses consistently. It would then be possible to search through the inhabitants of Ayr to find speakers of this variety, if any; but even if some were located, the problem of how to characterize the speech of those who were interviewed originally would remain. The search for a "pure" dialect/vernacular/standard is probably misguided. As John Rickford has noted (personal communication), there is no reason to believe that any variety of speech is monostylistic, since it will be used in a variety of situations, for a variety of ends, with different kinds of people. Rather than attempt to set up a set of rules by which different forms can be derived from a single underlying representation, it is probably more profitable to establish an inventory of forms from which speakers can select and attempt to determine the conditions under which the choice of a particular form is made. In other words, it is not the way in which [ʌu] and [u] are related (e.g., by Chomsky and Halle [1968]) that is important but the fact that a choice is available and also the conditions under which, for example, Gemmill chooses to say either *oot* or *out*.

The most satisfactory label for this inventory of forms and statement of conditions for their use is *dialect,* but it has to be clear that the term does not exclude variation. For example, Chambers and Trudgill, in attempting to explain the difference between accent and dialect, get trapped into taking an extreme position:

> "Accent" refers to the way in which a speaker pronounces, and therefore refers to a variety which is phonetically and/or phonologically different from other varieties. "Dialect," on the other hand, refers to varieties which are grammatically (and perhaps lexically) as well as phonologically different from other varieties. If two speakers say, respectively, *I done it last night* and *I did it last night,* we can say that they are speaking different dialects. (1980, 5)

As we have seen, speakers such as Lang use both *done* and *did* in this way in a single interview; and it seems less constructive to say that they are bidialectal than to argue that the dialect allows a choice between the two forms. A cautionary example on the other side comes from another introductory textbook when Petyt says: "So in a sense a dialect is an abstraction, based on some set of features chosen in a way which is essentially arbitrary: we have simply decided that we are going to take note of some features and ignore others when calling something 'a different form of a language' " (1980, 12). This is to place too much emphasis on the analyst and not enough on the language. The sense in which I want to use *dialect* is with reference to the speech of those who are indigenous to a locality, and there is nothing arbitrary about that, provided that the locality has some coherent identity. The problem arises when analysts try to group smaller geographical entities into larger units on the basis of isoglosses and label the speech of those larger units dialects. The difficulties in establishing firm boundaries of this kind have been well illustrated by Chambers and Trudgill in their examination of a transitional zone between northern forms and southern forms, in which they discovered "mixed lects" and "fudged lects" representative of neither the North or the South (1980, 125–42). Variation of this kind at spatial boundaries is probably the norm unless there is some psychological reality to the boundary. Speitel

(1969) and Glauser (1974) have shown that there is a greater bundling of isoglosses at the border between England and Scotland than for a considerable distance on either side of the border. It does not seem arbitrary that a significant division in language should coincide with a difference in national identification. In such situations language differences clearly serve an "identity maintenance function" (Giles, Scherer, and Taylor 1979, 358). Clear perceptual differences between communities are not restricted to national borders. One of the long-range goals of dialect geography must be to identify language boundaries that have arisen as the result of the perception of groups of individuals as to their shared characteristics (the unifying function of language) and their distinctness from other groups (the separatist function of language); but the study of dialects should not depend upon the identification of such boundaries. As Saussure pointed out a long time ago, "In the matter of language, people have always been satisfied with ill-defined units" (1966, 111).

I want to make the claim that the 12 speakers whose interviews form the basis of this work all speak "the dialect of Ayr." No claims are made as to how far "the dialect of Ayr" differs from other dialects in Ayrshire or the West of Scotland or beyond, since such a claim would require similar evidence from the other dialects for comparison and very little of this evidence exists at present. The speakers are judged to speak "the dialect of Ayr" on the grounds of their membership in the community and the belief that they are not perceived as outsiders by other members of the community. Since a dialect in this sense is multilayered (Kloss 1986, 101), the diversity of speech in the interviews does not reflect differences between dialects but differences within the dialect.

Klein (1988, 147) claims that he can recognize a speaker of Berlin dialect "after a few words" but admits that it is difficult to identify the linguistic features that create the "flavor" of the dialect. He goes on to speculate that "its specific flavour may arise from features hardly ever mentioned and surely never carefully studied in the literature, such as speech rate, pause structure or pitch range" (p. 147). I have had similar experiences overhearing Scottish speakers abroad. Even before I have understood any words or even identified critical phonetic segments, I have frequently recognized that the speaker was from the West of Scotland. It is, however, difficult to know how to investigate such features, since the methodology is complex and problematic (O'Connell 1988; Shapley 1989); and, as Klein points out, it is not even obvious what to look for. All the Ayr interviews, despite the range of speech on the tapes, have the characteristic Ayr "flavor" that reveals membership in the community.

To speak the dialect does not mean that one has to control the totality that makes up the dialect. It is unlikely that all 12 speakers knew, for example, the words *boyne, plab,* and *skeich* used by Rae; but the *potentiality* of knowing them is what constitutes the dialect. The kinds of lower-class features that middle-class speakers can mimic (Chapter 11) illustrate the kind of knowledge that exists. There are many other dialects in which this potentiality would not exist. To quote Saussure again: "It [langue] is a storehouse filled by the members of a community through their active use of speaking, a grammatical system that has a potential existence in each brain, or, more specifically, in the brains of a group of individuals. For language is not complete in any speaker; it exists perfectly only within a collectivity" (1966, 13–14). The key factor here is the potentiality that exists for members of the community "through their active use of speaking." Gemmill and Muir represent in many ways different ends of the social and linguistic continua; yet, as Muir remarked, they could talk to each other. This does not mean that either could imitate the other's speech or state with confidence what was grammatical in the other's form of speech, but the potentiality for both exists. It is unlikely that any of the middle-class speakers would be able to sustain a lower-class form of speech consistently over an extended exchange; but some could do so effectively in certain circumstances, and they would certainly understand what was said. In this they are similar to Dorian's semispeakers of East Sutherland Gaelic and near-passive bilinguals who are nevertheless accepted as members of the Gaelic speech community (Dorian 1981, 1982). Again, it is necessary to emphasize that even receptive knowledge of the dialect will not be uniform among the speakers; but the possibility of the knowledge exists for all members of the community. It is this potentiality which constitutes the speech community, though individuals will differ in the extent to which they take advantage of it. As Keith Lehrer has observed: "The primary contribution of coherence to a theory of language is to articulate what constitutes a communal language when, in fact, the people belonging to the community often express themselves in quite idiosyncratic idiolects. Beneath the surface of eccentric and idiosyncratic speech, coherence lies hidden" (1984, 43).

The evidence presented earlier suggests that the dialect of Ayr is a highly focused linguistic system in Le Page's sense of the term (Le Page and Tabouret-Keller 1985, 181–82). The continuity with Wilson's (1923) account of pronunciation demonstrated in Chapter 4 shows great stability. This is particularly significant in the case of the variable (a), where a similar range of allophony in Belfast has been seen as the result of dialect mixture (Trudgill 1986, 125). Ayr is very different from Belfast in that its growth has been moderate and steady and it is not torn by sectarian divisions. The different frequency with which the lower-class speakers use allophonic variants of (a) (or any of the other variables) is not the result of geographical location or network connections, since five of the six live within a few hundred yards of each other and have the opportunity to see each other regularly (the exception is Gemmill). It is not a particularly close network (though it is multiplex); but there is no reason to believe that the kinds of differences revealed in the interviews are because of other closer network ties, though there is no way to test this view from the evidence of the interviews alone.

What seems to be clear from the interviews is that all the lower-class speakers have a stylistic choice available to them in a number of variables that is not available in the same way to the middle-class speakers. Lang, in explaining how there is no conflict between Catholics and Protestants in Ayr, says:

(1)   WL2961   my best pal's a Catholic
                [seven lines omitted]
      2969   I'd say to him
      2970   "Well thae fellows are only *daeing* a job
      2971   aren't they?
      2972   the minister's *daeing* a job
      2973   and the priest's *daeing* a job"
      2974   I mean that's
      2975   how I look at it
      2976   it's a job
      2977   that they're *doing*

This section comes right at the end of the interview and is one of the rare occasions on which Lang uses *doing,* so there is no reason to believe that its occurrence is an accommodation to the interviewer or the interview situation. Instead, it is probably a mildly emphatic form triggered by the exposed position caused by the clefting. This is similar to the emphatic and metaphorical uses of (au) cited in Chapter 4.

If a middle-class speaker were to use *daeing* or similar forms outside of quoted direct speech, the effect would be much more marked. More likely in middle-class speech are what Aitken calls "overt Scotticisms" used by middle-class speakers to assert their Scottishness (1979, 106–10). For example, Nicoll is describing how he would make an exception for children:

(2)   DN169   "You let my wee boy
      170   play the harp"
      171   there's a harp there
      172   which is not permitted
      173   but you *cannae* say no to *bairns*

The lower-class form for children in the interviews is *weans;* but Nicoll uses the much older word *bairns,* which is a more usual "overt Scotticism," probably because it is not in the West of Scotland associated with lower-class speech. He also uses the clitic *–nae* in *cannae*. This is one of the few occasions on which he uses this form outside of quoted direct speech and is obviously intended to reinforce the effect of *bairns*.

The evidence from 12 interviews is a very slender basis on which to make strong claims, and the absence of younger speakers from the sample is particularly unfortunate; but the limited nature of the material has permitted a more intensive analysis than I would probably have undertaken if the sample had been larger. Although not all the items examined proved to be of equal interest, the presentation of a wide spectrum of evidence is intended to provide a basis for comparison with other varieties and future accounts of Ayr speech (though, judging by the comparison with Wilson's, there are not likely to be great changes in the immediate future). All those interviewed expressed their loyalty to the place. Lang's attitude is typical:

(3)   (Did you ever think of leaving?)
    WL2175   no never ever gave it a thought
    2176   never thought
    2177   of leaving this place at all
    2178   because I think
    2179   it's one of the cleanest wee toons in Scotland
    2180   oh aye Ayr's the tidiest wee toon in Scotland

Since the inhabitants of Ayr like their town the way it is, it is hardly surprising that they should manifest the same attitude toward their speech.

14

Conclusion

As was pointed out in the introduction, the present work is not the result of a carefully planned project to achieve this end. At the time I was collecting the interviews, I believed that I was continuing the kind of investigation I had carried out in Glasgow (Macaulay 1977), following the example of Labov (1966a). It was only when I began to look more closely at some of the interviews I had recorded in Ayr that I recognized the opportunity for a different kinds of analysis that they provided. I now believe that the kinds of results presented in the preceding chapters have important implications for sociolinguistic methodology.

During the past 20 years two distinct methods have been used to collect samples of speech for sociolinguistic surveys. One is the individual interview with a sample chosen in some nonsubjective manner (Labov 1966a; Shuy, Wolfram, and Riley 1968; Trudgill 1974). The other is to collect samples of interactive discourse under a predetermined set of conditions (Labov et al. 1968; Milroy 1980; Coupland 1988). While both methods have been fruitful and added to the knowledge of language variation, I want to emphasize here the disadvantages of both approaches as exemplified in the works cited.

The first method has been dominated by "the sociolinguistic interview" (Labov 1981, 7–9), which is designed to elicit a wide range of demographic, attitudinal, and social interaction information in addition to recording a variety of styles of speech and subjective reactions to examples of the speech of others. Probably, this was too ambitious a goal for a single interview; and the design of the questionnaire continually reinforced the notion of the interaction as "an interview." Since only the investigator actually listens to the tapes, it is impossible for readers of the published accounts to know how successful the interviews were in eliciting extended samples of speech; but the limited kinds of features that have been analyzed suggest certain limitations in the data.

The second method requires extensive planning and the simultaneous cooperation of a number of people. In Belfast, Lesley Milroy used network contacts to gain access to people's homes, where substantial amounts of impromptu speech were recorded. She makes it clear that powerful constraints affected the conditions under which the recording could be carried out (1980, 44). Labov and his coworkers spent a considerable amount of time and effort setting up the situation in which it was possible to record group sessions with adolescent gang members in New York (Labov et al. 1968). Coupland (1980) recorded one assistant in a travel agency in Cardiff, dealing with clients, answering the telephone, and in conversation with her fellow assistants, showing how her speech varied according to the situation and the person she was addressing. These are three very innovative and successful approaches to the study of linguistic variation, which apparently succeeded in recording remarkably varied samples of speech; but I (rather unkindly) wish to point out here the limitations of such approaches for a general sociolinguistic methodology. Each requires a special set of circumstances that may not be reproducible elsewhere, and this makes it difficult to obtain equivalent samples for comparative purposes. For example, Milroy could not have recorded similar samples of speech from middle-class households because the patterns of social interaction are very different. Likewise, though there may be some adult groups that it would be possible to record in situations similar to those in which Labov and his colleagues recorded the adolescent gangs, it would be difficult if not impossible set up such arrangements to record speakers from all sectors of the community. Coupland's study runs into the problems in the use of "confederates" in interactional research listed by Duncan and Fiske (1985, 6–12), because of the impossibility of controlling the variance.

While innovative approaches such as those of Lesley Milroy, Labov and his colleagues, and Coupland will always be valuable for the kind of data they record, the Ayr interviews analyzed in the present volume suggest that individual interviews (despite Wolfson 1976; Milroy and Milroy 1977; and Briggs 1986) can also provide a useful source of information on linguistic variation under more favorable conditions than those created by "the sociolinguistic interview." The advantage is that speech samples can be collected from a wide variety of speakers in a relatively short time. At this point, it may be helpful to repeat that the Ayr interviews were all with strangers—I had not known them before the recording took place. There was thus not the limitation caused by using only a small number of acquaintances (as, for example, in the studies by Douglas-Cowie [1978] and Broeck [1977]). Even as a stranger in town I was able to record the speech of individuals from widely varying backgrounds. More importantly, the interviews do not bear the signs of failure that Wolfson found to be characteristics of the so-called "spontaneous interview," namely, that "it has no rules of speaking to guide the subject or the interviewer" (1976, 195; see Chapter 1 above). The preceding analysis should have convincingly refuted Wolfson's claim that the speakers have "no rules of speaking" to guide them. On the contrary, the speakers display a wide variety of everyday conversational skills. They use discourse markers, introduce direct speech, initiate topics, perform narratives, and maintain both a global coherence and local cohesion in their speech. They reminisce, tell stories, express opinions, and ask questions. There is no sign that the speakers feel themselves to be incompetent in this situation. Although in some interviews I felt greater rapport with the speakers than in others, in general the result was what Albris and his colleagues (1988, 70) describe as "two intimate strangers talking together."

I believe that the kind of interviews collected in Ayr are a good example of the kind of speech samples that can be obtained in a "one-shot" approach to a complex speech community and could be carried out in widely differing situations. However, I have no doubt that some of the success of the interviews comes from the fact that I knew how to talk to people of this kind.[1] As Gal has observed: "Pragmatic analyses of the interview as a speech event suggest it is the ethnographer's task to discover the conditions under which informants can talk" (1989, 20). Thus, effort needs to be made in setting up the interviews—not in artificial devices intended to elicit different "styles of speech" (an effort that seems premature in the absence of any clear understanding of the notion *style*). Since the isolation of contextual styles in the interview is highly questionable anyway, the value of the sociolinguistic interview in its present form (Labov 1981) is diminished. Nor is there any compelling reason for the interviewer to ask the same questions in roughly the same order. Apart from essential background information, there is no question that the interviewer *must* ask, though questions about childhood, schooling, first job, and (often) marriage may lead to extended turns by the respondent. The rule, however, should be to observe the norms of polite conversation between strangers. (Even the elicitation of demographic information can be tricky, as when a question about father's occupation leads to an admission of illegitmacy.)[2]

---

[1]There are also important factors such as the age and gender of the interviewer and of the respondents. See Macafee 1988 for an illuminating discussion of the problems.

[2]Macafee (1988) includes a very interesting discussion of the advantages and disadvantages of group interviews. For her purpose (which was to collect information on the knowledge and use of local vocabulary) there were many advantages to group interviews, because people would prompt each other. However, she found that asking for personal details was a delicate matter because "people may see each other regularly without knowing, or wanting to know, much of each other's business" (p. 82). I found the same embarrassment in such a simple question as asking people's

As Duncan and Fiske observe with reference to a very different kind of research into face-to-face communication, "To go at something in interaction—an action, a message, a participant—as a separate entity, analyzable in itself, is not to court serious error, but virtually to ensure it" (1985, xviii). Duncan and Fiske refer to their work as "a voyage of discovery" (p. xix), and I have felt the same about the present study. To some readers it may seem that there is too much detail, to others too little; no doubt most readers would have liked more attention to this aspect and less to that. Duncan and Fiske quote Kendon on transcription: "No transcription, no matter how fine-grained, can ever be complete. One must inevitably make a selection" (1982, 479). Similarly, with analysis there must be some selection. The present work is not offered as a model for future studies but rather as an example of what can be done if the interviews are not treated as inert corpses from which pieces may be excised for examination and comparison with "specimens" from other sources. We need to develop not only a methodology for collecting sociolinguistic data but also a set of goals and established practices for analyzing the data. For example, in the present analysis, I have dealt solely with the language of the respondents; but obviously, my own speech behavior was a significant factor in the interaction. Whether much would have been gained by examining the interviewer's speech in detail is a question that cannot be answered at this time, but it would be worth investigating and would be crucial if there were more than one interviewer involved in the survey. Similarly, no systematic attention has been paid to prosodic and paralinguistic features, though these are probably as important as any of the features that were investigated (Klein 1988). Until there is some consensus on what we ought to be looking for and how we can obtain evidence for it, there is little hope that sociolinguistic investigation will progress successfully.

Meanwhile, it is rather ambitious in a sociolinguistic investigation to attempt to describe how a speech variety is changing in the absence of an adequate description of the variety as it is (or has been) spoken. We may have been trying to run before we can walk, and such a form of locomotion can be sustained only for a short time before the inevitable fall. It might be helpful if the empirical basis for linguistic theories were a little stronger.

The analysis in the present work has attempted to provide a wide range of information on the speech of "honest men" and (fewer) "bonnie lassies" in Ayr. In the absence of comparable studies, it is impossible to know to what extent the results are typical of (1) the speakers, (2) the community, (3) Scottish English speakers, or (4) English speakers. (With so many unknowns

---

age in a group situation. Macafee has also some very interesting observations on the problem of a woman's interviewing men.

in the sampling procedures, statistical measures of significance are more or less irrelevant.) However, enough information has been given to provide a basis for comparison with future studies here or elsewhere.

No attempt has been made to draw theoretical conclusions from the mass of data. To some this will merely confirm their suspicions that the mouse that the mountain has been laboring so long to produce is ridiculously small, but I will be content if it is considered to be well formed. In Virginia Woolf's *To the Lighthouse,* Lily Briscoe ponders the meaning of life: "What is the meaning of life? That was all—a simple question; one that tended to close in on one with years. The great revelation had never come. The great revelation perhaps never did come. Instead there were little daily miracles, illuminations, matches struck unexpectedly in the dark; here was one" (1938, 186). I hope that for some of those interested in the mystery of human language the present work will prove to be a match struck unexpectedly in the dark.

APPENDIX I

Number of Each Syntactic Unit in Individual Interviews

	Lower-Class Respondents[a]						
Type[b]	EL	HG	WR	WL	MR	AS	All
ca	299	357	71	412	92	553	1,784
cb	94	116	32	102	47	161	552
cf	0	6	0	0	1	0	7
co	5	5	4	5	0	9	28
cs	60	44	18	75	3	25	225
ds	285	46	56	286	25	114	812
fr	20	57	98	16	33	20	244
ge	40	28	8	81	24	125	302
im	10	0	4	4	0	4	22
in	114	69	6	88	40	280	597
ma	894	444	304	704	300	821	3,467
mc	8	14	0	25	4	6	57
md	2	1	4	6	0	10	23
me	4	5	1	9	2	17	38
mf	22	15	11	29	6	32	115
ml	6	23	4	6	1	16	56
mo	22	25	27	37	9	12	132
mp	4	0	2	0	0	1	7
mr	7	6	7	12	2	3	37
mt	15	52	24	46	27	33	197
mv	39	32	29	48	15	16	179

(continued)

Lower-Class Respondents[a] (*Continued*)

Type[b]	EL	HG	WR	WL	MR	AS	All
my	4	0	7	12	0	7	30
nw	46	54	8	22	10	82	222
pp	41	29	6	38	12	40	388
qu	61	11	17	20	5	13	127
rn	0	12	0	0	0	11	23
ro	46	38	17	46	12	79	238
rs	3	7	0	2	1	7	20
rt	21	52	5	25	15	112	230
rw	4	0	0	1	0	26	31
sa	24	7	6	8	6	93	144
sb	39	16	10	12	6	52	135
sc	1	4	2	7	2	7	23
sd	7	0	3	2	3	14	29
si	41	25	17	25	11	39	158
sm	14	3	1	9	2	13	42
sn	78	53	7	58	23	183	402
so	2	1	0	0	0	0	3
sp	18	7	2	17	0	27	71
sq	16	22	6	18	9	36	107
sr	2	9	1	1	0	3	16
ss	0	0	0	0	0	3	3
st	91	59	20	80	34	152	436
sw	0	1	0	12	0	2	15
z	21	5	21	32	15	14	108
zq	24	53	76	49	78	47	327

Middle-Class Respondents[c]

	DN	AM	NM	JM	IM	WG	All
ca	539	223	147	228	106	174	1,417
cb	137	71	39	59	44	77	427
cf	0	0	0	0	0	0	0
co	7	8	1	5	4	1	26
cs	43	10	6	13	10	50	132
ds	434	6	2	47	7	158	654
fr	28	24	40	28	13	25	158
ge	53	76	23	32	18	63	265
im	9	10	2	0	1	0	22
in	181	190	38	81	56	108	654

Middle-Class Respondents[c] (*Continued*)

Type[b]	DN	AM	NM	JM	IM	WG	All
ma	605	267	301	287	269	396	2,125
mc	2	4	0	1	1	6	14
md	0	0	0	0	0	0	0
me	7	4	2	5	2	5	25
mf	21	12	7	19	8	16	83
ml	3	1	1	0	0	2	7
mo	43	6	14	4	12	11	90
mp	0	0	0	0	0	0	0
mr	0	0	0	1	0	0	1
mt	20	13	17	10	4	27	91
mv	76	8	29	15	21	16	165
my	4	1	0	3	1	0	9
nw	22	12	6	5	14	16	75
pp	48	25	4	17	5	32	131
qu	27	16	21	9	14	12	99
rn	35	3	4	10	13	19	84
ro	26	11	9	16	5	20	87
rs	11	5	5	3	4	6	34
rt	18	36	13	9	3	33	112
rw	27	30	11	31	10	32	141
sa	35	8	8	10	7	21	89
sb	26	67	14	13	21	23	164
sc	0	16	1	2	4	8	31
sd	4	7	0	2	2	4	19
si	42	41	12	31	11	27	164
sm	3	0	2	1	2	5	13
sn	101	193	62	72	82	109	619
so	0	0	0	0	0	0	0
sp	13	9	2	7	1	6	38
sq	33	25	15	12	10	10	105
sr	2	16	1	4	0	5	28
ss	0	1	0	1	0	1	3
st	48	50	22	54	47	26	247
sw	0	1	0	1	0	0	2
z	69	52	12	8	20	31	192
zq	11	51	61	45	49	46	263

[a]EL = Laidlaw, HG = Gemmill, WE = Rae, WL = Lang, MR = Ritchie, AS = Sinclair.

[b]For these abbreviations, see Chapter 3.

[c]DN = Nicoll, AM = MacDougall, NM = Menzies, JM = MacGregor, IM = Muir, WG = Gibson.

APPENDIX II

Topic Introduction

The following examples illustrate how the speakers introduce topics in the interviews:

(1)   *Overt introductions*

DN2648	uh two little stories about that
2661	another little story about that
EL1823	I'll tell you a story aboot him
HG364	and I'll tell you what I did
1604	I'll tell an experience I had
WR112	well to gie you an idea
AS1778	oh to give you the best example
2488	I mean I'll give you a good example
494	I'll give you a good wee example of it

(2)   *Cataphoric reference*

WG933	what one finds of course
DN173	one of the things I treasure
HG235	see the biggest problem in this was
1239	this was the thing
WL1016	this is how it's worked
AS2188	but he really is—this is—he's a
2957	you see this comprehensive education

(3)   *Recollection*

JM121	I'll never forget
NM285	and I remember
EL512	and I remember once
WR303	I mind o yin instance
623	oh but this I dae remember
WL1046	I mind a man coming to see me
AS1369	I remember as a boy

(4)   *Reference to past time*

WG355	for a long time I was
1363	I later met
AM654	at that time I—I tried
1537	my father used to
NM1239	and in winter
JM1600	but in those days
DN856	many years later ah there was
1469	one day at C—— we had
EL157	he was at one time
706	if I made a mistake
2351	but when she was young
2685	I was about forty-year married the first time and
HG527	in nineteen-twenty-six there was a miner's strike
714	and another day
WL1339	the last day I was at school
2682	it wouldnae be six-months old when
2825	and yet the minister at that time was
AS1737	and during that time there

(5)   *Reference to habitual speech*

HG1664	and—they talk aboot the good old days
WL210	and eh as I say
1708	I used to say
2784	but as I say
AS83	I mean as you say
WG756	I often say
844	that's why I say
DN609	but as I often say to people

(6)   *Epistemic reference*

WG1503	I think this is
AM1172	in fact I think
1190	certainly I think
1575	yeah I was quite—I was wondering
IM747	I don't want to get talking politics but I think
JM1813	I know that
WL2236	I think the last year I was there
AS453	I think this was
2744	and I don't know whether I'm right in saying this

(7)   *Anaphoric reference*

WG383	and that's something that
JM897	I only had that job
DN497	uh BJ did exactly the same as you

1329	possibly you heard me
2648	uh two little stories about that
EL1078	we never had anything like that
1419	you saw thon two
1468	but that's no the point
2037	they did exactly the same with
2186	so he is
MR72	that's what I missed
HG828	the thing aboot it was
WL1252	this was through Mr H
1873	and that big fellow was doon there
2428	and that's where I kept the banties
EL1754	she's been unlucky that way
1858	he belongs to Fraserburgh
2081	Uncle Johnny paid it
HG1265	but wunst they were auld
WL1461	and he was a great man
WG357	I was talking about that the other day
1300	in (. . .) that I mentioned earlier
DN463	and as I said
1832	I was chatting with a family from Ireland earlier in the morning
EL2657	ah we were speaking to
HG771	I was telling

(8)   *Additional reference*

DN2661	another lovely story about
EL1546	the other lassie
HG595	and the other point was
661	I had another experience of this
WL522	there another wee man
742	and there another wee one doon at Dailly
AS2136	and there was another wee fellow

(9)   *Exemplification*

WG493	an instance was
1027	I get a thrill for instance
WR490	I mind o yin instance
AS1778	oh to give you the best example
2488	I mean I'll give you a good example
AS3494	I'll give you a good wee example of it

(10)   *Conjunctions*

NM1505	and a crate of apples used to be a good thing
EL1770	and I've a blind daughter—in Edinburgh
2274	and he died just [snaps fingers]

WR490	and the farmers—that was the yin thing
WL968	and noo again fae I left the pits
2306	and we'd the night oot in Blackpool
AS2075	and John used to get on the bus
	[67 examples of *and*]
WG474	but I find this even now
JM527	but you know we had—
DN609	but as I say often to people
EL2523	but the vandalism's noo—it's ridiculous
WL265	but oh there's great changes
984	but mind you I don't know
AS1936	but eh to tell you some of the things
	[40 examples of *but*]
DN1141	so uh having got round to the academy
EL2186	so he is
2365	so I goes oot this night
HG337	so when I was sixty

(11)   *Discourse markers*

WG519	well we'll give parts
DN2536	well we all have a favorite thing
EL1554	well you see these kind are out of homes
HG320	well I've been retired noo for aboot twelve year
WR112	well to gie you an idea
WL302	well there are a great difference
1629	well with him daeing it
AS719	well I went away
1766	well I say I've settled doon noo
HG1386	oh photos in there I've hunners of them
WR623	oh but this I dae remember
WL265	but oh there's great changes in the pits
1190	oh aye I used to go with a baund tae
1298	oh aye and it was—
1413	oh and it was a great thing this
1649	oh he was some manager
2626	oh he was the best
2729	but oh the tears that we shed
2769	oh I wore the leggings as a boy
AS1177	oh it was a great experience
1778	oh to give you the best example
2121	oh he wrote one thing aboot mince and tatties
HG235	see the biggest problem in this was
WL1101	see that's what I get
1672	see there was a sort of a field

352	aye you see it wasnae
431	you see I sat my papers
943	you see the pit places doon there
1593	aye you see I had a brother
AS2957	you see this comprehensive education
3106	you see I've only three of a family
AM204	actually I'm quite interested in
1172	in fact I think
MR604	in fact there soon—
NM1155	and of course so many wives
WR513	coorse in the Miners Rows it was
IM359	now my son—em—I've sent both my sons
EL896	now my grandson is six
1513	now that girl that you saw
DN556	funny I've had someone from America
HG887	anyway she go a—kinna—not a very good end
AS083	I mean as you say Whitletts
2488	I mean I'll give you a good example

(12) *Agreement markers*

AM1575	yeah I was quite—I was wondering
1602	yeah I presume
DN2123	yes we were talking the story about
EL2146	aye he was good
WL352	aye you see it wasnae
1190	oh aye I used to go with a baund tae
1298	oh aye and it was
1593	aye you see I had a brother
2140	aye and we won a trip to Miss World
2341	aye I've a girl works in Slough
2385	aye I was doon there
2498	aye and I kept ferrets

(13) *Exophoric reference*

EL1049	see when that goes on (*that* = TV set)
WL1101	see that's what I get (*that* = grandchildren squabbling)
2368	he's some dug that (referring to the dog that had growled)

See Macaulay 1989b for a discussion of topic initiation in the interviews.

References

Abercrombie, David. 1965. *Studies in Phonetics and Linguistics*. London: Oxford University Press.

——. 1979. "The Accents of Standard English in Scotland." In *Languages of Scotland*, ed. Adam J. A. Aitken and Tom McArthur, 68–84. Edinburgh: Chambers.

Aitken, Adam J. 1979. "Scottish Speech: A Historical View with Special Reference to the Standard English of Scotland." In *Languages of Scotland*, ed. Adam J. Aitken and Tom McArthur, 85–115. Edinburgh: Chambers.

——. 1981. "The Scottish Vowel-Length Rule." In *So Meny People Longages and Tonges*, ed. Michael Benskin and Michael L. Samuels, 131–57. Edinburgh: Middle English Dialect Project.

——. 1985. "A History of Scots." In *The Concise Scots Dictionary*, ed. Mairi Robinson, ix–xx. Aberdeen: Aberdeen University Press.

Albris, Jon, Frans Gregersen, Henrik Holmberg, Erik Møller, Inge Lise Pedersen, and Ole Nedergaard Thomsen. 1988. *The Copenhagen Study in Urban Sociolinguistics: Interim Report*. Copenhagen: Institute of Nordic Philology.

Arena, Louis A. 1982. "The Language of Corporate Attorneys." In *Linguistics and the Professions*, ed. Robert J. Di Pietro, 143–54. Norwood, NJ: Ablex.

Bakhtin, Mikhail. 1973. *Problems of Dostoevsky's Poetics*. Trans. R. W. Rotsel. Ann Arbor.

Bates, E., M. Masling, and W. Kintsch. 1978. "Recognition Memory for Aspects of Dialogue." *Journal of Experimental Psychology: Human Learning and Memory* 4:187–97.

Bell, A. 1984. "Language Style As Audience Design." *Language in Society* 13:145–204.

Berdan, Robert. 1978. "Multidimensional Analysis of Vowel Variation." In *Linguistic variation: Models and methods*, ed. David Sankoff, 149–60. New York: Academic.

Bernstein, Basil. 1962. "Social Class, Linguistic Codes, and Grammatical Elements." *Language and Speech* 5:221–40.

——. 1971. *Class, Codes, and Control*, Vol. 1. London: Routledge & Kegan Paul.

Bever, Thomas G. 1972. "Perceptions, Thought, and Language." In *Language Comprehension and the Acquisition of Knowledge*, ed. Roy O. Freedle and John B. Carroll, 99–112. Washington: Winston.

Biber, Douglas. 1986. "Spoken and Written Textual Dimensions in English." *Language* 62:384–414.

———. 1988. *Variation Across Speech and Writing*. Cambridge: Cambridge University Press.

Biber, Douglas, and Edward Finegan. 1988. "Adverbial Stance Types in English." *Discourse Processes* 11:1–34.

Bickerton, Derek. 1973. "The Nature of a Creole Continuum." *Language* 49:640–69.

Bloomfield, Leonard. 1927. "Review of The Philosophy of Grammar by Otto Jespersen," *Journal of English and Germanic Philology* 26:444–46.

Boissevain, Jeremy. 1974. *Friends of Friends: Networks, Manipulators, and Coalitions*. Oxford: Blackwell.

Bower, Anne R. 1983. "The Construction of Stance in Conflict Narrative." Ph.D. diss., University of Pennsylvania.

Brenneis, Donald. 1986. "Shared Territory: Audience, Indirection, and Meaning." *Text* 6: 339–47.

Briggs, Charles L. 1986. *Learning How To Ask: A Sociolinguistic Appraisal of the Role of the Interview in Social Science Research*. Cambridge: Cambridge University Press.

Broeck, Jef van den. 1977. "Class Differences in Syntactic Complexity in the Flemish Town of Maaseik." *Language in Society* 6:149–81.

———. 1984. "Why Do Some People Speak in a More Complicated Way Than Others?" In *Sociolinguistics in the Low Countries*, ed. Kas Deprez, 45–68. Amsterdam: Benjamins.

Brown, Gillian, and George Yule. 1983. *Discourse Analysis*. Cambridge: Cambridge University Press.

Brown, Penelope, and Stephen C. Levinson. 1987. *Politeness*. Cambridge: Cambridge University Press.

Chafe, Wallace L., ed. 1980. *The Pear Stories: Cognitive, Cultural, and Linguistic Aspects of Narrative Production*. Norwood, NJ: Ablex.

———. 1986a. "How We Know Things about Language: A Plea for Catholicism." In *Languages and Linguistics: The Interdependence of Theory, Data, and Application*, ed. Deborah Tannen and James E. Alatis, 214–25. Washington: Georgetown University Press.

———. 1986b. "Beyond Bartlett: Narratives and Remembering." *Poetics* 15:139–51.

———. 1987. "Evidentiality in English Conversation and Academic Writing." In *Evidentiality: The Linguistic Coding of Epistemology*, ed. Wallace L. Chafe and Johanna Nichols, 261–72. Norwood, NJ: Ablex.

Chambers, J. K., and Peter Trudgill. 1980. *Dialectology*. Cambridge: Cambridge University Press.

Cheshire, Jenny. 1978. "Present Tense Verbs in Reading English." In *Sociolinguistic Patterns in British English*, ed. Peter Trudgill, 52–68. London: Arnold.

———. 1982. *Variation in an English Dialect*. Cambridge: Cambridge University Press.

Chomsky, Noam, and Morris Halle. 1968. *The Sound Pattern of English*. New York: Harper & Row.

Cichocki, Wladyslaw. 1988. "Uses of Dual Scaling in Social Dialectology: Multidimensional Analysis of Vowel Variation." In *Methods in Dialectology*, ed. Alan R. Thomas, 187–99. Clevedon: Multilingual Matters.

Coombs, Clyde H. 1964. *A Theory of Data*. New York: Wiley.

Coulmas, Florian, ed. 1986. *Direct and Indirect Speech*. Berlin: Mouton de Gruyter.

Coupland, Nikolas. 1980. "Style-Shifting in a Cardiff Work Setting." *Language in Society* 9:1–12.

———. 1988. *Dialect in Use: Sociolinguistic Variation in Cardiff English*. Cardiff: University of Wales Press.

Craen, Pete van de. 1984. "Social Linguistics and the Failure of Theories." In *Sociolinguistics in the Low Countries*, ed. Kas Deprez, 103–28. Amsterdam: Benjamins.

Dines, Elizabeth R. 1980. "Variation in Discourse—'and Stuff Like That'." *Language in Society* 9:13–31.

Dorian, Nancy C. 1981. *Language Death: The Life-Cycle of a Scottish Gaelic Dialect*. Philadelphia: University of Pennsylvania Press.

———. 1982. "Defining the Speech Community to Include Its Working Margins." In *Sociolinguistic Variation in Speech Communities*, ed. Suzanne Romaine, 25–33. London: Arnold.

Douglas-Cowie, E. 1978. "Linguistic Code-Switching in a Northern Irish Village: Social Interaction and Social Ambition." In *Sociolinguistic Patterns in British English*, ed. Peter Trudgill, 37–51. London: Arnold.

Du Bois, John W. 1980. "Beyond Definiteness: The Trace of Identity." In *The Pear Stories*, ed. Wallace L. Chafe, 203–74. Norwood, NJ: Ablex.

———. 1987. "The Discourse Basis of Ergativity." *Language* 63: 805–55.

Duncan, Starkey, Jr., and Donald W. Fiske. 1985. *Interaction Structure and Strategy*. Cambridge: Cambridge University Press.

Duranti, Alessandro. 1984. "The Social Meaning of Subject Pronouns in Italian Conversation." *Text* 4:277–311.

Feagin, Crawford. 1979. *Variation and Change in Alabama English: A Sociolinguistic Study of the White Community*. Washington: Georgetown University Press.

Gal, Susan. 1989. "Between Speech and Silence: The Problematics of Research on Language and Gender." *Papers in Pragmatics* 3 (no. 1): 1–38.

Giles, Howard, Klaus R. Scherer, and Ronald M. Taylor. 1979. "Speech Markers in Social Interaction." In *Social Markers in Speech*, ed. Klaus R. Scherer and Howard Giles, 343–81. Cambridge: Cambridge University Press.

Giles, Howard, and Peter M. Smith. 1979. "Accommodation Theory: Optimal Levels of Convergence." In *Language and Social Psychology*, ed. Howard Giles and Robert St. Clair, 45–65. Oxford: Blackwell.

Givón, Talmy. 1979a. *On Understanding Grammar*. New York: Academic.

———, ed. 1979b. *Discourse and Semantics*. Syntax and Semantics, vol. 12. New York: Academic.

———, ed. 1983. *Topic Continuity in Discourse: A Quantitative Cross-Linguistic Study*. Amsterdam: Benjamins.

Glassie, Henry. 1982. *Passing the Time in Ballymenone*. Philadelphia: University of Pennsylvania Press.

Glauser, Beat. 1974. *The Scottish–English Linguistic Border: Lexical Aspects*. Bern: Francke Verlag.

Goffman, Erving. 1974. *Frame Analysis*. New York: Harper & Row.

———. 1981. *Forms of Talk*. Philadelphia: University of Pennsylvania Press.

Goldman-Eisler, Frieda. 1968. *Psycholinguistics: Experiments in Spontaneous Speech*. New York: Academic.

Goodwin, Margaret H. 1982. "'Instigating': Storytelling As Social Process." *American Ethnologist* 9:799–819.

Grant, William. 1913. *The Pronunciation of English in Scotland*. Cambridge: Cambridge University Press.

Grant, William and James M. Dixon. 1921. *Manual of Modern Scots*. Cambridge: Cambridge University Press.

Gregg, Robert J. 1958. "Notes on the Phonology of a County Antrim Scotch–Irish Dialect." *Orbis* 7:392–406.

Grice, H. P. 1975. "Logic and Conversation." In *Speech Acts,* ed. Peter Cole and Jerrold Morgan, 41–58. New York: Academic.

Grimes, Joseph E. 1975. *The Thread of Discourse*. The Hague: Mouton.

Gumperz, John J. 1968. "The Speech Community." In *International Encyclopedia of the Social Sciences,* ed. by David L. Sills, 381–86. New York: Macmillan.

———. 1971. *Language in Social Groups*. Stanford: Stanford University Press.

———. 1982. *Discourse Strategies*. Cambridge: Cambridge University Press.

Haiman, John. 1985. *Natural Syntax: Iconicity and Erosion*. Cambridge: Cambridge University Press.

Halliday, Michael A. K. 1967. *Intonation and Grammar in British English*. The Hague: Mouton.

———. 1985. *Introduction to Functional Grammar*. London: Arnold.

———. 1987. "Spoken and Written Modes of Meaning." In *Comprehending Oral and Written Language,* ed. Rosalind Horowitz and S. Jay Keyser, 55–82. San Diego: Academic.

Haugen, Einar. 1980. "How Should a Dialect be Written?" In *Dialekt und Dialektologie,* eds. Joachim Göschel, Pavle Ivić, and Kurt Kehr. *Zeitschrift für Dialektologie und Linguistik,* vol. 26, 273–82. Wiesbaden: Steiner.

Heath, Christopher D. 1980. *The Pronunciation of English in Cannock, Staffordshire*. Oxford: Philological Society.

Hinds, John. 1979. "Organization Patterns in Discourse." In *Syntax and Semantics,* vol. 12, *Discourse and Syntax,* ed. Talmy Givon, 35–57. New York: Academic.

Hjelmquist, Erland. 1984. "Memory for Conversations." *Discourse Processes* 7:321–36.

Hopper, Paul J. 1979. "Aspect and Foregrounding in Discourse." In *Discourse and Semantics,* ed. Talmy Givón, 213–41. New York: Academic.

Horvath, Barbara. 1985. *Variation in Australian English*. Cambridge: Cambridge

University Press.

———. 1987. "Text in Conversation: Variability in Story-Telling Texts." In *Variation in Language:NWAV-XV at Stanford,* eds. Keith M. Denning, Sharon Inkelas, Faye C. McNair-Knox, and John R. Rickford, 212–23. Stanford: Department of Linguistics, Stanford University.

Hudson, Richard A. 1980. *Sociolinguistics.* Cambridge: Cambridge University Press.

Hunt, Kellogg W. 1965. *Grammatical Structures Written at Three Grade Levels.* Champaign, IL: National Council of Teachers of English.

Hurtig, Richard. 1977. "Toward a Functional Theory of Discourse." In *Discourse Production and Comprehension,* ed. Roy O. Freedle, 89–106. Norwood, NJ: Ablex.

Huspek, Michael. 1989. "Linguistic Variability and Power: An Analysis of YOU KNOW/I THINK Variation in Working-Class Speech." *Journal of Pragmatics* 13:661–83.

Hymes, Dell H. 1972. "Models of the Interaction of Language and Social Life." In *Directions in Sociolinguistics,* ed. John J. Gumperz and Dell Hymes, 35–71. New York: Holt, Rinehart, and Winston.

———. 1974. *Foundations in Sociolinguistics.* Philadelphia: University of Pennsylvania Press.

———. 1984. "Sociolinguistics: Stability and Consolidation." *International Journal of the Sociology of Language* 45:39–45.

———. 1986. "Discourse: Scope without Depth." *International Journal of the Sociology of Language* 57:49–89.

Irvine, Judith. 1979. "Formality and Informality in Communicative Events." *American Anthropologist* 81:773–90.

Jefferson, Gail. 1978. "Sequential Aspects of Story-telling." In *Studies in the Organization of Conversational Interaction,* ed. Jim Schenkein, 219–48. New York: Academic.

Johansson, Stig, and Knut Hofland. 1989. *Frequency Analysis of English Vocabulary and Grammar.* 2 vols. Oxford: Oxford University Press.

Johnstone, Barbara. 1987. "'He Says . . . So I Said': Verb Tense Alternation and Narrative Depictions of Authority in American English." *Linguistics* 15:33–52.

———. 1988. "Individual Variation in Discourse Marking." Paper presented at the Linguistic Society of America Annual Meeting, New Orleans.

Keenan, Edward L. 1975. "Variation in Universal Grammar." In *Analyzing Variation in Language,* ed. Ralph Fasold and Roger W. Shuy, 136–49. Washington: Georgetown University Press.

Keenan, Edward L., and Bernard Comrie. 1977. "Noun Phrase Accessibility and Universal Grammar." *Linguistic Inquiry* 8:63–99.

Keenan, Elinor O., and Bambi B. Schieffelin. 1976. "Topic As a Discourse Notion: A Study of Topic in the Conversation of Children and Adults." In *Subject and Topic,* ed. Charles N. Li, 335–84. New York: Academic.

Kendon, Adam. 1982. "The Organization of Behavior in Face-to-Face Interaction:

Observations on the Development of a Methodology." In *Handbook of Methods in Nonverbal Behavior Research,* ed. Klaus R. Scherer and Paul Ekman, 441–505. Cambridge: Cambridge University Press.

Klein, Wolfgang. 1988. "The Unity of a Vernacular: Some Remarks on 'Berliner Stadtsprache'." In *The Sociolinguistics of Urban Vernaculars: Case Studies and Their Evaluation,* ed. Norbert Dittmar and Peter Schlobinski, 147–53. de Gruyter.

Kloss, Heinz. 1986. "On Some Terminological Problems in Interlingual Sociolinguistics." *International Journal of the Sociology of Language* 57:91–106.

Kroch, Anthony, and C. Small. 1978. "Grammatical Ideology and Its Effect on Speech." In *Linguistic Variation: Models and Methods,* ed. David Sankoff, 45–56. New York: Academic.

Labov, William. 1963. "The Social Motivation of a Sound Change." *Word* 19:273–309.

———. 1966a. *The Social Stratification of English in New York City.* Washington: Center for Applied Linguistics.

———. 1966b. "Hypercorrection by the Lower Middle Class As a Factor in Linguistic Change." In *Sociolinguistics,* ed. William Bright, 84–113. The Hague: Mouton.

———. 1969. "Contraction, Deletion, and Inherent Variability of the English Copula." *Language* 45:715–62.

———. 1972. *Language in the Inner City.* Philadelphia: University of Pennsylvania Press.

———. 1981. "Field Methods of the Project on Linguistic Change and Variation." Sociolinguistic Working Paper No. 81. Southwest Educational Development Laboratory (Austin).

———. 1982. "Building on Empirical Foundations." In *Perspectives on Historical Linguistics,* ed. Winfred P. Lehmann and Yakov Malkiel, 79–92. Amsterdam: John Benjamins.

———. 1984. "Intensity." In *Meaning, Form, and Use in Context: Linguistic Applications,* ed. Deborah Schiffrin, 43–70. Washington: Georgetown University Press.

Labov, William, Paul Cohen, Clarence Robins, and John Lewis. 1968. *A Study of the Non-standard English of Negro and Puerto Rican Speakers in New York City.* Cooperative Research Report No. 3288. New York: Columbia University.

Lakoff, George. 1972. "Hedges: A Study in Meaning Criteria and the Logic of Fuzzy Concepts." In *Papers from the Eighth Regional Meeting of the Chicago Linguistic Circle,* eds. Paul M. Peranteau, Judith N. Levi, and Gloria C. Phares, 83–228. Chicago: Chicago Linguistic Society.

———. 1987. *Women, Fire, and Dangerous Things.* Chicago: University of Chicago Press.

Lakoff, Robin. 1973. "Language and Woman's Place." *Language in Society* 2:45–80.

Lavandera, Beatriz R. 1978. "Where Does the Sociolinguistic Variable Stop?" *Lan-*

guage in Society 7:171–82.

Lawton, Denis. 1968. *Social Class, Language, and Education.* London: Routledge & Kegan Paul.

Lee, David. 1987. "The Semantics of *Just." Journal of Pragmatics* 11:377–98.

Lehrer, Keith. 1984. "Coherence, Consensus, and Language." *Linguistics and Philosophy* 7:43–55.

Le Page, R. B. and Andrée Tabouret-Keller. 1985. *Acts of Identity: Creole-based Approaches to Language and Ethnicity.* Cambridge: Cambridge University Press.

Levinson, Stephen C. 1983. *Pragmatics.* Cambridge: Cambridge University Press.

———. 1988. "Conceptual Problems in the Study of Regional and Cultural Style." In *The Sociolinguistics of Urban Vernaculars,* ed. Norbert Dittmar and Peter Schlobinski, 161–90. Berlin: de Gruyter.

Lindenfeld, Jacqueline. 1969. "The Social Conditioning of Syntactic Variation in French." *American Anthropologist* 71:890–98.

Local, John. 1986. "Patterns and Problems in a Study of Tyneside Intonation." In *Intonation in Discourse,* ed. Catherine Johns-Lewis, 181–98. London: Croom Helm.

Macafee, Caroline. 1980. "Characteristics of Non-Standard Grammar in Scotland." University of Glasgow. Typescript.

———. 1988. "Some Studies in the Glasgow Vernacular." Ph.D. diss., University of Glasgow.

McAllister, Anne H. 1963. *A Year's Course in Speech Training.* London: University of London Press.

Macaulay, Ronald K. S. 1975a. "Negative Prestige, Linguistic Insecurity, and Linguistic Self-Hatred." *Lingua* 36:145–61.

———. 1975b. "Linguistic Insecurity. In *The Scots Language in Education,* ed. J. Derrick McClure, 35–43. Aberdeen: Aberdeen College of Education.

———. 1976a. "Social Class and Language in Glasgow." *Language in Society* 5:173–88.

———. 1976b. "Tongue-Tied in the Scottish Classroom." *Education in the North* 13:13–16.

———. 1977. *Language, Social Class, and Education: A Glasgow Study.* Edinburgh: Edinburgh University Press.

———. 1978a. "Variation and Consistency in Glaswegian English." In *Sociolinguistic Patterns in British English,* ed. Peter Trudgill, 132–43. London: Arnold.

———. 1978b. "The Myth of Female Superiority in Language," *Journal of Child Language* 5:353–63.

———. 1984. "Chattering, Nattering, and Blethering: Informal Interviews As Speech Events." In *Studies in Linguistic Ecology,* ed. Werner Enninger and Lilith Haynes, 51–64. Wiesbaden: Franz Steiner Verlag.

———. 1985. "The Narrative Skills of a Scottish Coal Miner." In *Focus on: Scotland,* ed. Manfred Görlach, 101–24. Amsterdam: John Benjamins.

———. 1986. "Measures of Syntactic Complexity." Paper presented at the New Ways of Analyzing Variation in English XV conference, Stanford University.

———. 1987a. "The Sociolinguistic Significance of Scottish Dialect Humor." *International Journal of the Sociology of Language* 65:53–63.

———. 1987b. "Polyphonic Monologues: Quoted Direct Speech in Oral Narratives." *IPRA Papers in Pragmatics* 1:1–34.

———. 1988a. "Linguistic Change and Stability." In *Linguistic Change and Contact*, eds. Kathleen Ferrara, Becky Brown, Keith Walters, and John Baugh, 225–31. Austin: University of Texas.

———. 1988b. "A Microsociolinguistic Study of the Dialect of Ayr." In *Methods in Dialectology*, ed. Alan R. Thomas, 456–63. Philadelphia: Multilingual Matters.

———. 1988c. "The Rise and Fall of the Vernacular." In *On Language: Rhetorica, Phonologica, Syntactica*, ed. Caroline Duncan-Rose and Theo Vennemann, 106–13. London: Routledge.

———. 1988d. "RP R. I. P." *Applied Linguistics* 9:115–24.

———. 1989a. "'He Was Some Man Him': Emphatic Pronouns in Scottish English." In *Synchronic and Diachronic Approaches to Linguistic Variation and Change*, ed. Thomas J. Walsh, 179–87. Washington: Georgetown University Press.

———. 1989b. "Changing the Topic." Paper presented at NWAVE XVIII, Duke University.

———. 1991. "'Coz It Izny Spelt When They Say It': Displaying Dialect in Writing." Forthcoming in *American Speech*.

Macaulay, Ronald K. S., and Gavin D. Trevelyan. 1973. *Language, Education, and Employment in Glasgow*. Report to the Social Science Council. Edinburgh: Scottish Council for Research in Education.

Malinowski, Brontislav. 1935. *Coral Gardens and Their Magic*. London: Allen & Unwin.

Mather, James Y., and Hans-Henning Speitel. 1975–86. *The Linguistic Atlas of Scotland*. 3 vols. London: Croom Helm.

Miller, George A. 1951. *Language and Communication*. New York: McGraw-Hill.

———. 1956. "The Magical Number Seven Plus or Minus Two: Some Limits on Our Capacity for Processing Information." *Psychological Review* 63:81–97.

Milroy, James. 1981. *Regional Accents of English: Belfast*. Belfast: Blackstaff.

———. 1982. "Probing under the Tip of the Iceberg: Phonological 'Normalisation' and the Shape of Speech Communities." In *Sociolinguistic Variation in Speech Communities*, ed. Suzanne Romaine, 35–47. London: Arnold.

Milroy, James, and Lesley Milroy. 1977. "Speech and Context in an Urban Setting." *Belfast Working Papers in Language and Linguistics* 2:1–85.

Milroy, Lesley. 1980. *Language and Social Networks*. Oxford: Blackwell.

———. 1987. *Observing and Analysing Natural Language: A Critical Account of Sociolinguistic Method*. Oxford: Blackwell.

Mitchell-Kernan, Claudia. 1972. "Signifying and Marking: Two Afro–American Speech Acts." In *Directions in Sociolinguistics*, ed. John J. Gumperz and Dell

Hymes, 161–79. New York: Holt.

Moore, Terence, and Christine Carling. 1982. *Understanding Language: Towards a Post-Chomskyan Linguistics.* London: Macmillan.

Ochs, Elinor. 1979. "Transcription As Theory." In *Developmental Pragmatics,* ed. Elinor Ochs and Bambi B. Schieffelin, 43–72. New York: Academic.

O'Connell, Daniel C. 1988. *Critical Essays on Language Use and Psychology.* New York: Springer-Verlag.

Orton, Harold. 1962. *Survey of English Dialects: Introduction.* Leeds: Arnold.

Östman, Jan-Ola. 1981. *You Know: A Discourse-Functional Approach.* Amsterdam: John Benjamins.

———. 1982. "The Symbiotic Relationship between Pragmatic Particles and Impromptu Speech." In *Impromptu Speech: A Symposium,* ed. Nils Erik Enkvist, 147–77. Abo: Abo Akademi.

Pawley, Andrew, and Frances Syder. 1977. "The One-Clause-at-a-Time Hypothesis. University of Auckland. Typescript.

Petyt, K. M. 1980. *The Study of Dialect: An Introduction to Dialectology.* London: Deutsch.

———. 1982. "Who Is Really Doing Dialectology?" In *Linguistic Controversies,* ed. David Crystal, 50–72. London: Arnold.

Polanyi, Livia. 1982. "Literary Complexity in Everyday Storytelling." In *Spoken and Written Language: Exploring Orality and Literacy,* ed. Deborah Tannen, 155–70. Norwood, NJ: Ablex.

———. 1985. "Conversational Storytelling." In *Handbook of Discourse Analysis,* vol. 3, ed. Teun van Dijk, 183–201. London: Academic.

Pomerantz, Anita. 1978. "Compliment Responses: Notes on the Cooperation of Multiple Constraints." In *Studies in the Organization of Conversational Interaction,* ed. Jim Scheinkein. New York: Academic.

———. 1984. "Agreeing and Disagreeing with Assessments: Some Features of Preferred/Dispreferred Turn Shapes." In *Structures of Social Action: Studies in Conversational Analysis,* ed. J. Maxwell Atkinson and John Heritage, 57–101. Cambridge: Cambridge University Press.

Prince, Ellen. 1981. "Toward a Taxonomy of Given–New Information." In *Radical Pragmatics,* ed. Peter Cole, 223–55. New York: Academic.

Quirk, Randolph, Sidney Greenbaum, Geoffrey Leech, and Jan Svartvik. 1972. *A Grammar of Contemporary English.* London: Longman.

———. 1985. *A Comprehensive Grammar of the English Language.* London: Longman.

Reichman, Rachel. 1978. "Conversational Coherency." *Cognitive Science* 3:283–327.

———. 1985. *Getting Computers to Talk Like You and Me: Discourse Context, Focus, and Semantics (an ATN Model).* Cambridge: MIT Press.

Reid, Euan. 1978. "Social and Stylistic Variation in the Speech of Children: Some Evidence from Edinburgh." In *Sociolinguistic Patterns in British English,* ed. Peter Trudgill, 158–71. London: Arnold.

Robinson, Mairi. 1985. *The Concise Scots Dictionary.* Aberdeen: Aberdeen University Press.

Romaine, Suzanne. 1975. "Linguistic Variability in the Speech of Some Edinburgh Schoolchildren." Master's thesis. University of Edinburgh.

———. 1978. *A Sociolinguistic Investigation of Edinburgh Speech.* Report to the Social Science Research Council.

———. 1980a. "A Critical Overview of the Methodology of Urban British Sociolinguistics." *English World-Wide* 1:163–98.

———. 1980b. "The Relative Clause Marker in Scots English: Diffusion, Complexity, and Style As Dimensions of Syntactic Change." *Language in Society* 9:221–49.

———. 1981. "Syntactic Complexity, Relativization, and Stylistic Levels in Middle Scots." *Folia linguistica historica* 2:56–77.

———. 1982a. *Socio–Historical Linguistics: Its Status and Methodology.* Cambridge: Cambridge University Press.

———, ed. 1982b. *Sociolinguistic Variation in Speech Communities.* London: Arnold.

———. 1984. *The Language of Children and Adolescents: The Acquisition of Communicative Competence.* Oxford: Blackwell.

Rosenbaum, Harvey. 1979. "Relative Clause Structures in Informal Conversation." In *Studies Presented to Emmon Bach by His Students,* ed. Elisabeth Engdahl and Mark J. Stein, 165–73. Amherst: University of Massachusetts.

Rydén, Mats. 1983. "The Emergence of *Who* As Relativizer." *Studia linguistica* 37:126–34.

Sacks, Harvey, Emanuel Schegloff, and Gail Jefferson. 1974. "A Simplest Systematics for the Organization of Turn-Taking for Conversation." *Language* 50:696–735.

Sankoff, Gillian. 1980. *The Social Life of Language.* Philadelphia: University of Pennsylvania Press.

Sankoff, Gillian, and D. Vincent. 1977. "L'emploi productif du *ne* dans le français parlé à Montreal." *Le Français moderne* 34:81–108.

Saussure, Ferdinand de. 1922. *Cours de linguistique générale.* 2d ed. Paris: Payot.

———. 1966. *Course in General Linguistics.* Trans. Wade Baskin. New York: McGraw-Hill.

Schegloff, Emanuel A., and Harvey Sacks. 1973. "Opening Up Closings." *Semiotica* 8:289–327.

Schiffrin, Deborah. 1980. "Meta-Talk: Organizational and Evaluative Brackets in Discourse." *Sociological Inquiry* 50:199–234.

———. 1981. "Tense Variation in Narrative." *Language* 57:45–62.

———. 1983. "Discourse Markers: Pragmatic Coordinators of Talk." University of Pennsylvania. Typescript.

———. 1985. "Conversational Coherence: The Role of *Well.*" *Language* 61:640–67.

———. 1987. *Discourse Markers.* Cambridge: Cambridge University Press.

Schourup, Lawrence C. 1985. *Common Discourse Particles in English Conversation.* New York: Garland.

Searle, John R. 1969. *Speech Acts.* Cambridge: Cambridge University Press.

Shapley, Marian. 1989. "Fundamental Frequency Variation in Conversational Discourse." Ph.D. diss., University of California, Los Angeles.

Shuy, Roger W., Walter A. Wolfram, and William K. Riley. 1968. *Field Techniques in an Urban Language Study*. Washington: Center for Applied Linguistics.

Speitel, Hans-Henning. 1969. "A Typology of Isoglosses: Isoglosses near the Scottish–English Border." *Zeitschrift für Dialektologie und Linguistik* 7:49–66.

Stenström, Anna-Brita. 1987. "What Does *Really* Really Do?" In *Grammar in the Construction of Texts*, ed. James Monaghan, 65–79. London: Pinter.

Strawhorn, John. 1975. *Ayrshire: The Story of a County*. Ayr: Ayrshire Archaelogical and Natural History Society.

Svartvik, Jan, and Randolph Quirk (eds.) 1980. *A Corpus of English Conversation*. Lund: Gleerup.

Tannen, Deborah, ed. 1982. *Spoken and Written Language: Exploring Orality and Literacy*. Norwood, NJ: Ablex.

———, ed. 1984. *Coherence in Spoken and Written Discourse*. Norwood, NJ: Ablex.

———. 1986. "Introducing Constructed Dialogue in Greek and American Conversational and Literary Narrative." In *Direct and Indirect Speech*, ed. Florian Coulmas, 311–32. Berlin: Mouton de Gruyter.

———. 1987. "Repetition and Variation As Spontaneous Formulaicity in Conversation." *Language* 63:574–605.

———. 1989. *Talking Voices: Repetition, Dialogue, and Imagery in Conversational Discourse*. Cambridge: Cambridge University Press.

Thakerar, Jitendra N., Howard Giles, and Jenny Cheshire. 1982. "Psychological and Linguistic Parameters of Speech Accommodation Theory." In *Advances in the Social Psychology of Language*, ed. Colin Fraser and Klaus R. Scherer, 205–55. Cambridge: Cambridge University Press.

Thompson, Sandra A. 1983. "Grammar and Discourse: The English Detached Participial Clause." In *Discourse Perspectives on Syntax*, ed. Flora Klein-Andreu, 43–65. New York: Academic.

———. 1984. "'Subordination' in Formal and Informal Discourse." In *Meaning, Form, and Use in Context: Linguistic Applications*, ed. Deborah Schiffrin, 85–94. Washington: Georgetown University Press.

———. 1987. "'Subordination' and Narrative Event Structure." In *Coherence and Grounding in Discourse*, ed. Russell S. Tomlin, 435–54. Amsterdam: John Benjamins.

———. 1988. "'*That*-Deletion': A Discourse Perspective." Typescript.

Traugott, Elizabeth, and Suzanne Romaine. 1985. "Some Questions for the Definition of "Style" in Sociohistorical Linguistics." *Folia Linguistica* 6:7–39.

Trudgill, Peter. 1974. *The Social Differentiation of English in Norwich*. Cambridge: Cambridge University Press.

———. 1982. "On the Limits of Passive 'Competence': Sociolinguistics and the Polylectal Grammar Controversy." In *Linguistic Controversies*, ed. David Crystal, 172–91. London: Arnold.

———. 1986. *Dialects in Contact*. Oxford: Blackwell.

Trudgill, Peter, Viv K. Edwards, and Bert Weltens. 1983. *Grammatical Charac-*

teristics of Non-Standard Dialects of English in the British Isles: A Survey of Research. Report to the Social Science Research Council (London).

Van Dijk, Teun A. 1982. "Episodes As Units of Discourse Analysis." In *Analyzing Discourse: Text and Talk,* ed. Deborah Tannen, 177–95. Washington: Georgetown University Press.

Venneman, Theo. 1975. "Topic, Sentence Accent, and Ellipsis: A Proposal for Their Formal Treatment." In *Formal Semantics of Natural Language,* ed. Edward L. Keenan, 313–28. Cambridge: Cambridge University Press.

Voloshinov, V. N. 1986. (Orig. publ. 1929) *Marxism and the Philosophy of Language.* Trans. Ladislav Matejka and I. R. Titunik. Cambridge, Mass.: Harvard University Press.

Weiner, E. J., and William Labov. 1983. "Constraints on the Agentless Passive." *Journal of Linguistics* 19:29–58.

Wells, J. C. 1982. *Accents of English.* Cambridge: Cambridge University Press.

Wilson, James. 1923. *The Dialect of Robert Burns As Spoken in Central Ayrshire.* Oxford: Oxford University Press.

———. 1926. *The Dialects of Central Scotland.* Oxford: Oxford University Press.

Wolfram, Walt. 1969. *A Sociolinguistic Description of Detroit Negro Speech.* Washington: Center for Applied Linguistics.

Wolfson, Nessa. 1976. "Speech Events and Natural Speech: Some Implications for Sociolinguistic Methodology." *Language in Society* 5:189–211.

———. 1978. "A Feature of Performed Narrative: The Conversational Historical Present." *Language in Society* 7:215–39.

———. 1982. *CHP: The Conversational Historical Present in American English Narrative.* Dordrecht: Foris.

Woolf, Virginia. 1938. *To the Lighthouse.* London: Dent. (Everyman Library)

Wright, Peter. 1972. "Coal-Mining Language: A Recent Investigation." In *Patterns in the Folk Speech of the British Isles,* ed. Martyn F. Wakelin, 32–49. London: Athlone.

———. 1974. *The Language of British Industry.* London: Macmillan.

Index